T0319539

Python for Geospatial Data Analysis

Theory, Tools, and Practice
for Location Intelligence

Bonny P. McClain

Beijing · Boston · Farnham · Sebastopol · Tokyo

Python for Geospatial Data Analysis

by Bonny P. McClain

Published by O'Reilly Media, Inc., 1005 Gravenstein Highway North, Sebastopol, CA 95472.

O'Reilly books may be purchased for educational, business, or sales promotional use. Online editions are also available for most titles (*http://oreilly.com*). For more information, contact our corporate/institutional sales department: 800-998-9938 or *corporate@oreilly.com*.

Acquisitions Editor: Michelle Smith	**Indexer:** Ellen Troutman-Zaig
Development Editor: Sarah Grey	**Interior Designer:** David Futato
Production Editor: Kate Galloway	**Cover Designer:** Karen Montgomery
Copyeditor: Shannon Turlington	**Illustrator:** Kate Dullea
Proofreader: Tom Sullivan	

October 2022:   First Edition

Revision History for the First Edition

2022-10-19:   First Release

See *http://oreilly.com/catalog/errata.csp?isbn=9781098104795* for release details.

978-1-098-10479-5

[LSI]

Table of Contents

Preface

Python for Geospatial Data Analysis is an introduction to geospatial data analysis using a selection of Python libraries and packages, optimized for exploration and discovery. In recent years, as the Internet of Things has arisen and location data has become available on local computers, professionals in a wide range of fields have been exploring geospatial platforms with built-in analytics—geospatial professionals, data scientists, business analysts, geographers, geologists, researchers, analysts, computer scientists, and statistics professionals, to name just a few. As they explore, these learners often seek to gain a deeper appreciation of the technology embedded in their tools. Graphical user interfaces are powerful, but you can truly elevate your skill set by writing your own Python code to fully customize or automate the outcome and by gaining a better grasp of how large platforms and systems work.

Open source projects look to engage a wide variety of users; today, professionals in most industries likely have access to location data and publicly available datasets. Advances in cloud computing no longer require massive downloads to local computers, meaning that access has been democratized to anyone with an internet-enabled device. This book aims to be a resource that meets professionals wherever they are and serves as a guide to their goals and understanding. This book encourages readers of all backgrounds to engage in geospatial analysis to help inform decision making in urban planning, climate change research, and many other fields.

Even for experienced data analysts, technology is often siloed. You may know a bit of Python and can jump in and out of QGIS or ArcGIS without really needing to know operationalized workflows. You can certainly work with embedded systems without appreciating raster, vector, or mathematical models, but understanding these foundational elements will bring a new level of rigor and inquiry to your exploratory and explanatory data competency.

How did I end up writing a book about geospatial analytics and Python integration? I had a problem to solve. In conducting analysis on electronic health records, I found that they incorporated plenty of medical data but only vaguely useful demographic information. Examining this aspect of the built infrastructure of health care revealed to me the importance of location in predicting health outcomes. We can't make policy decisions about public health without knowing the answers to questions like:

- Do patients' neighborhoods offer fresh markets within walking distance or along accessible transportation lines?
- Are patients' neighborhoods walkable and safe?
- Do patients have access to green spaces?
- How does proximity to highways, power lines, and waste-treatment plants, for example, influence the health of nearby neighborhoods?
- And how can we bring location data into studying these environments?

Integrating these types of nonspatial data with spatial information allows you to explore your surroundings—or any other place on the Earth's surface that interests you—and generate maps and other visualizations that can help you think through complex questions.

Why Python?

Python is a popular, nimble scripting language. Its accessible coding syntax makes it easier to learn than most programming languages, so you can start using it relatively quickly. Python is great for geospatial data analysis because it has already been incorporated into many geographic information system (GIS) applications, including ArcGIS and QGIS. It also has the support of a vibrant open source community with an abundance of libraries and packages.

This is *not* a Python 101 book. If you're a beginner looking for a basic grounding in the language, there are many resources available to you. You might want to start with *Pandas 1.x Cookbook,* second edition, by Matt Harrison and Theodore Petrou (Packt Publishing), or any of the in-person or online courses by Dunder Data, such as *Master the Fundamentals of Python* (*https://oreil.ly/swed0*).

Perhaps you don't have the inclination or time to learn an entire programming language. Although I recommend having at least one computer language in your toolbox, that's OK. The purpose of this book is to get you up and running with the extensive ecosystem of publicly available geospatial data, and total programming competence is not required. I provide step-by-step guidance along with the code snippets throughout the book. I direct you to resources on a variety of additional functions not covered in this book so that you can take what you learn and use it to explore. As you become comfortable working with one or two functions in a large library or package, you will gain confidence to expand into new areas.

How This Book Works

We will begin with a quick level set introducing you to a few key GIS concepts. As we advance, I'll slowly integrate Python learning. I do not assume expertise in a coding language or in geospatial analytics.

The resources presented here are open source: their source code is distributed freely by the developers and usually incorporates contributions from community members. Most make use of Python. To the best of my ability, I've ensured that this book features resources available without burdensome subscription services. Any costs, however modest, are highlighted so that you can make informed decisions. My focus on open source doesn't mean that I don't support enterprise solutions; it means I want to lower the barriers to conducting meaningful analysis around big questions.

This book covers a wide swath of open source tools and data and looks at a variety of datasets, some of which you perhaps don't have access to in your current professional role. *Python for Geospatial Data Analysis* is not linear like typical books about technology (or, for that matter, about Python). There are multiple ways to explore data problems. Perhaps you can draw inspiration from working in an integrated development environment (IDE) for the first time. Maybe you're curious about working in a terminal or console.

It is impossible to walk through the granular details of each Python package or library in a single book. If you're familiar with a particular tool or library, you probably have favorite features that I haven't included here. That's fine—I just want to give you a feel for each one. From there, you can continue discovering what they have to offer.

Who Is This Book For?

My vision for *Python for Geospatial Data Analysis* presented me with a conundrum: how do you write a book for newly minted geospatial professionals who know Python *and* for newly minted Python programmers who are well versed in geospatial analytics? I decided simply to make it interesting. My goal isn't to grant you professional expertise at either end of the spectrum but to bring us all together to learn tools and best practices.

By the end of this book, I want all of you to feel proficient and confident enough to go out and explore geospatial analytics on your own. To that end, as I teach each tool and technique, I ask you to follow along, installing the tools as needed and using a Jupyter or Google Colab notebook to run code. But I don't want you to stop there, so I also provide a host of different experiences that invite you to continue to explore.

A Few Tips on Tooling

There is no such thing as infallible code. You are going to need to learn how to troubleshoot. That is why the data goddess invented Stack Overflow (*https://stackover flow.com*). A quick warning—you don't always get the best or most correct solutions from advice sites, so research with caution. To help, I have selected Python packages with a solid GitHub presence to ensure that support is always an option. Having said that, I'll offer a few tips you may find helpful as you install and get acquainted with the tools in this book:

- Create Python environments when possible so that you can control versions and dependencies.

- If you aren't in a Conda or Mamba environment, always check your versions of Python and any other software or packages. Version issues are the biggest generator of error codes.

- Not sure if your pip install worked? Write `!pip list` in a cell and run it. All packages installed in that session will be included in the list.

- Don't be afraid of package documentation! Reading the instructional materials is an important part of acquiring new skills and is key to troubleshooting.

Finding Your Way

The first two chapters offer some general concepts and competencies you'll need as you dive into geospatial analytics. From there, we'll move into specific tools, using hands-on activities to help you get familiar with them. Chapter 3 covers QGIS, and Chapter 4 looks at Google Earth Engine and other cloud-based analytics tools. Chapter 5 introduces you to OpenStreetMap, Chapter 6 shows you around the ArcGIS Python API, and Chapter 7 explores spatial statistics with the GeoPandas library. Chapter 8 pauses for a look at data cleaning: a process for separating the useful data from the noise that often comes with it. Chapter 9 then introduces a phenomenally useful resource: the Geospatial Data Abstraction Library (GDAL). Finally, in Chapter 10, three hands-on projects bring all of your learning together to show how useful your geospatial analytics skills can be in researching a pressing global issue: climate change.

At the end of the book, you'll find a rich trove of links to tools, packages, and data resources as well as citations for the sources used in the text and recommended reading to help you continue your learning journey.

Conventions Used in This Book

The following typographical conventions are used in this book:

Italic
> Indicates new terms, URLs, email addresses, filenames, and file extensions.

`Constant width`
> Used for program listings, as well as within paragraphs to refer to program elements such as variable or function names, databases, data types, environment variables, statements, and keywords.

`Constant width bold`
> Shows commands or other text that should be typed literally by the user.

`Constant width italic`
> Shows text that should be replaced with user-supplied values or by values determined by context.

> This element signifies a tip or suggestion.

> This element signifies a general note.

> This element indicates a warning or caution.

Using Code Examples

Supplemental material (code examples, exercises, etc.) is available for download at *https://github.com/datamongerbonny/geopy-notebooks.git*.

If you have a technical question or a problem using the code examples, please send email to *bookquestions@oreilly.com*.

This book is here to help you get your job done. In general, if example code is offered with this book, you may use it in your programs and documentation. You do not need to contact us for permission unless you're reproducing a significant portion of the code. For example, writing a program that uses several chunks of code from this book does not require permission. Selling or distributing examples from O'Reilly books does require permission. Answering a question by citing this book and quoting example code does not require permission. Incorporating a significant amount of example code from this book into your product's documentation does require permission.

We appreciate, but generally do not require, attribution. An attribution usually includes the title, author, publisher, and ISBN. For example: "*Python for Geospatial Data Analysis* by Bonny P. McClain (O'Reilly). Copyright 2023 Grapheme Consulting, Inc., 978-0-098-10479-5."

If you feel your use of code examples falls outside fair use or the permission given above, feel free to contact us at *permissions@oreilly.com*.

O'Reilly Online Learning

 For more than 40 years, *O'Reilly Media* has provided technology and business training, knowledge, and insight to help companies succeed.

Our unique network of experts and innovators share their knowledge and expertise through books, articles, and our online learning platform. O'Reilly's online learning platform gives you on-demand access to live training courses, in-depth learning paths, interactive coding environments, and a vast collection of text and video from O'Reilly and 200+ other publishers. For more information, visit *https://oreilly.com*.

How to Contact Us

Please address comments and questions concerning this book to the publisher:

O'Reilly Media, Inc.
1005 Gravenstein Highway North
Sebastopol, CA 95472
800-998-9938 (in the United States or Canada)
707-829-0515 (international or local)
707-829-0104 (fax)

We have a web page for this book, where we list errata, examples, and any additional information. You can access this page at *https://oreil.ly/python-for-geo-data*.

Email *bookquestions@oreilly.com* to comment or ask technical questions about this book.

For news and information about our books and courses, visit *https://oreilly.com*.

Find us on LinkedIn: *https://linkedin.com/company/oreilly-media*.

Follow us on Twitter: *https://twitter.com/oreillymedia*.

Watch us on YouTube: *https://www.youtube.com/oreillymedia*.

Acknowledgments

I would like to thank the many geospatial students and professionals alike offering guidance, suggestions, and questions that were the motivation for tackling this topic. The enthusiasm for a book that is both a point of entry and useful guide for the experienced has been a source of great pride.

I am grateful for the tremendous work and support offered by Qiusheng Wu, Assistant Professor in the Department of Geography at the University of Tennessee, Knoxville. Dr. Wu's contributions to the geospatial community are unparalleled, and it was his presentation, Google Earth Engine and geemap workshop at GeoPython Conference 2021 (*https://oreil.ly/X4Pv6*), that introduced me to the important integration of Google Earth Engine and Python. One year later, I presented at the very same conference.

I also would like to thank Ted Petrou, masterful expert in Python and data exploration at Dunder Data (*https://www.dunderdata.com*). Ted offers just-in-time resources and affordable workshops to fill the gap between didactic exposure to Python and applying the skills in your work environment.

Ujaval Gandhi of SpatialThoughts (*https://oreil.ly/3eJ0c*) has been instrumental in introducing geospatial platforms and skills to both enrolled and independent learners. I was on the receiving end of his expertise and generosity on more than one occasion.

A heartfelt thank you and appreciation of camaraderie is extended to my Geospatial Connections (*https://oreil.ly/SeeNU*) community moderators: Bruce Buxton, Juliana McMillan-Wilhoit, Tim Nolan, and Kendrick Faison. The conversations in this space with our colleagues have been nothing short of amazing and informative.

I thank my husband, Steve, for keeping me supplied with good humor, endless support, big salads, and an inexhaustible supply of Gummy Bears. To my sons, Harrison and Ryland, thanks for being my North Stars, endless sources of wit and charm, and my reason for everything.

CHAPTER 1

Introduction to Geospatial Analytics

Are you a geographer, geologist, or computer scientist? Impressive, if you answered yes! I'm none of those: I am a spatial data analyst, interested in exploring data and integrating location information into data analysis.

Geospatial data is collected everywhere. Appreciating the *where* in data analyses introduces a new dimension: comprehending the impact of a wider variety of features on a particular observation or outcome. For instance, I spend a lot of professional time examining large open source datasets in public health and health care. Once you become familiar with geocoding and spatial files, not only can you curate insights across multiple domains, but you can also recognize and target areas where profound social and economic gaps exist.

Early in my evolution as a data analyst, I began to realize I had bigger and more complex questions to consider, and I needed more resources. With an eye toward working with United States census data, I enrolled in a course in applied analytics. I had worked in the R programming language, but this course was taught in Python. I made it through, but I discovered a lot in the following months that I wish I had learned along with basic Python. This book is meant to share what I wish I'd been taught.

What I hope to share here is not the complete coding paradigm of Python, nor is it a Python 101 course. Instead, it is meant to supplement your Python learning by showing you how to write actionable code so that you can learn by doing. The book contains simple examples that explore key concepts in detail. The graphics in earlier chapters will familiarize you with how the maps look and how different tools can render relationships. Later chapters explore the code and various platforms, empowering you to use Python as a resource for answering geospatial questions. If you are looking for flexibility in manipulating data in both open source and proprietary systems, Python may be the missing piece of the puzzle. It is fairly easy to learn and has a variety of libraries for pivoting and reshaping tables, merging data, and generating plots.

Integrating Python into spatial analysis is the focus of this book. Open source platforms like OpenStreetMap (OSM) (*http://openstreetmap.org*) allow us to zoom in and add attributes. OSMnx is a Python package that lets you download spatial data from OSM and model, project, visualize, and analyze real-world street networks and structures in the landscape in a Jupyter Notebook, independent of any specific application or tool.[1] You can download and model walkable, drivable, or bikeable urban networks with a single line of Python code, then easily analyze and visualize them. You can also download and work with other infrastructure types, amenities and points of interest, building footprints, elevation data, street bearings and orientations, speed, and travel times. Chapter 5 of this book offers an opportunity to dig a little deeper into OSM.

Using the principles of spatial data analysis, you can consider challenges at the local, regional, and global levels. These might include the environment, health care, biology, geography, economics, history, engineering, local government, urban planning, and supply chain management, to name a few. Even issues that seem local or regional cross physical and political boundaries, ecological regions, municipalities, and watersheds and possess a spatial component. Since maps are one of the first appreciations we have of data visualization, it makes sense that after interrogating your data you may become curious about location. This chapter will introduce you to some of the broad objectives of spatial data analysis and look at how geospatial information can affect our thinking.

Democratizing Data

The accessibility of open source data tools and massive open online courses (MOOCs) has empowered a new cohort of citizen scientists. Now that the general public has access to location data and geospatial datasets, many people are becoming "data curious," regardless of their professional titles or areas of study. Perhaps you're a bird watcher and you're interested in a certain species of bird—say, the blue heron. You might want to access spatial data to learn about its habitat. You might ask research questions like: Where are blue herons nesting? Where do they migrate? Which of their habitats support the most species, and how is this changing over time? You might create maps of your own sightings or other variables of interest.

---

1 Boeing, G. 2017. "OSMnx: New Methods for Acquiring, Constructing, Analyzing, and Visualizing Complex Street Networks." *Computers, Environment and Urban Systems* 65: 126–139. *https://doi.org/10.1016/j.compenvurbsys.2017.05.004.*

In addition to personal or professional hobbies, geospatial analysts examine the socioeconomics of neighborhoods and how they change geographically over time. The study of environmental racism seeks to analyze how the built infrastructure can inhibit or influence health outcomes in marginalized communities. We'll explore this idea a little later, when we generate a data question to explore.

The Map Warper (*https://mapwarper.net*) project is an open source collection of historical maps and current locations. The challenge is that older maps contain many errors due to outdated survey technology. The Map Warper project is a map-rectification project: it aims to correct those errors to match today's precise maps by searching for modern matching ground control points and warping the image accordingly. These control points are known coordinates spaced within areas of interest utilized as precise known locations. You can use rectified maps to explore development over time in different cities or to reimagine a historical location more accurately. How do investments in infrastructure and industrial development affect neighborhoods over time, for example? Figure 1-1 shows a rectified map. You can browse already rectified maps or assist the New York Public Library by aligning a map yourself. Everyone is welcome to participate!

There are many opportunities for professionals across multiple industries to include location intelligence in their analytics. Location intelligence is actionable information derived from exploring geospatial relationships—that is, formulating data questions and evaluating hypotheses. Since open source tools are welcoming new end users, we need a lexicon that works for people with diverse interests, resources, and learning backgrounds. I want you to be able to explore all of the tools this book discusses.

Although powerful subscription-based applications and tools are available, they are mostly priced as enterprise solutions, not for individual users, which limits access for anyone not affiliated with a large institution. Take GIS software, for example: there are many options, all with pros and cons. Two of the most prominent options are the Aeronautical Reconnaissance Coverage Geographic Information System, now known as ArcGIS, which I discuss in Chapter 6, and Quantum GIS, or QGIS, which is the topic of Chapter 3. In my professional work I use both, but when teaching, I like to give QGIS the main stage because it is truly open source: you don't need to pay for different levels of licensing to access its tools.

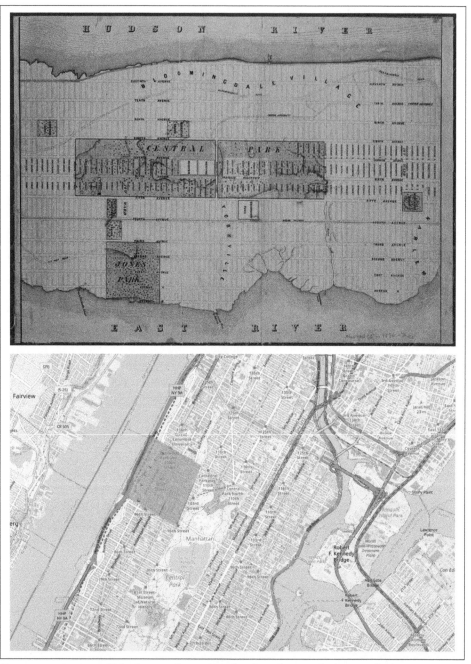

Figure 1-1. A map of Manhattan from 1870 (top) and a rectified map of contemporary Manhattan (bottom)

I learned the hard way just how expensive subscription-based tools can be. ArcGIS uses a credit system, and when I began using it, I didn't realize that the paid service would kick in automatically when I uploaded a CSV file with location data. Don't make that mistake! QGIS, on the other hand, offers two options—both free.

Asking Data Questions

National census bureaus are a rich source of data, especially about demographics. I generated the map of the United States shown in Figure 1-2 in ArcGIS using US census data: specifically, county-level race variables from the American Community Survey (ACS).[2] Publicly available resources like the US Census Bureau provide an *application programming interface (API)*: a data transmission interface that lets anyone retrieve demographic data with a few lines of Python code. You can pull the specific data you want instead of downloading an entire massive dataset.

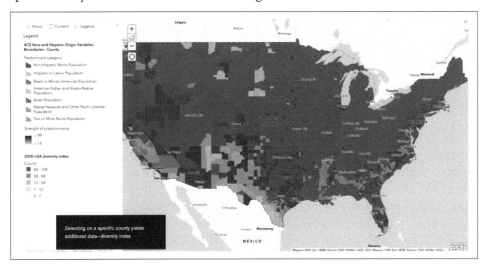

Figure 1-2. An ArcGIS map visualizing county-level race variables in the continental US

The polygons on the map are shaded in different colors corresponding to the racial categories used in the census and show where each group represents the majority of the local population. At first glance you can see clusters of categorical variables, but

---

2 The annual ACS replaced the long form of the decennial census in 2005. It asks a wide variety of questions to identify shifting demographics and gather information about local communities. Geographic census data from around the world is also available for download from the Integrated Public Use Microdata Series (IPUMS) (*https://www.ipums.org*). IPUMS's integration and documentation make it easy to study change, conduct comparative research, merge information across data types, and analyze individuals within family and community context. Its data and services are available free of charge.

what else might you want to know? This is where geospatial data becomes so valuable. Can you answer your question with the data you have? You may need to reformulate your question if faced with missing data or resources you are unable to access. What might you determine about these clusters if you also examine other attributes?

You will have the opportunity to add nonspatial data as an additional layer in Chapter 7, when we use the Census Data API to explore US census data. For now, I will give you a hint: the devil is in the details—or should I say, the layers. Layers are collections of geographic data you can select based on your specific query. For example, in the layer in Figure 1-2, polygons are shaded based on areas where a certain race is the majority population.

This is a good place to introduce the expression: the defaults should not be the endpoints. The default settings are where you should begin your analysis. What information is summarized in your data that might be best explored with additional granularity? A deeper understanding of GIS will allow you to move away from default settings and create unique and deeper insights.

You will explore how customizing your variable selection specifically for spatial and nonspatial attributes enhances the questions you ask and the insights you gather.

The first rule of spatial data analysis is that *any analysis requires a defined question*. What do you want to know? Do you have a hypothesis that you want to test? Once you formulate your research question, you can look for data that will help answer it.

When formulating a data question, I like to refer to Tobler's First Law of Geography: "Everything is related to everything else, but near things are more related than distant things."[3] Applied directly to geospatial concepts, when you think about objects being "near," the concept extends to time and space, not simply adjacency. For example, oceanfront homes may experience the direct impact of rising sea levels and intensified storms, but the ensuing flooding can disrupt a much larger region. You have to learn to think of spatial connectivity when thinking about spatial configurations and the "nearness" described in connected space.

To be clear, that doesn't mean cherry-picking data to match the answers you want. You need to look at all of the relevant data to shape a hypothesis or generate an insight. In thinking about how to understand race and racism in the US, you might seek to reveal policy gaps, address unmet health needs, or foster empathy. Working backward from your goal, you might use geospatial data to examine how race intersects with place in arenas such as housing, employment, transportation, and education, to name a few. When you become familiar with census data, you begin to understand the heavy lifting race has been asked to accomplish. That's the kind of

---

3 Tobler, W. 1970. "A Computer Movie Simulating Urban Growth in the Detroit Region." *Economic Geography* 46 (Supplement): 234–240. *https://doi.org/10.2307/143141.*

research that spurred me to integrate geospatial applications like ArcGIS and QGIS into my talks about poverty, racial inequity, structural determinants of health, and a wide variety of emerging questions.

When you rely on spreadsheets or tables without spatial data, you may be missing out on critical insights. Nonspatial data describes how values are distributed—and you can rely on descriptive statistics. But what if you are curious about the impact spatial relationships might have on these values? Static metrics, such as the location of a road or a specific event, as well as dynamic measures like the spread of an infectious disease become more powerful when you integrate them with location intelligence. *Spatial analysis* examines the relationships between features identified within a geographic boundary.

A Conceptual Framework for Spatial Data Science

Geospatial problems are complex and change over space and time. Look no further than current headlines to find examples: racial inequity, climate change, structural determinants of health, criminal justice, water pollution, unsustainable agriculture practices, ocean acidification, poverty, species endangerment and extinction, and economic strife. How does an individual's location influence their health, well-being, or economic opportunity? You can answer questions like these using GIS, by discovering and showing spatial patterns in phenomena such as the diffusion rates of diseases, patients' distances to the nearest hospital, and the locations of roadways, waterways, tree cover, and city walkability.

Spatial thinking includes considerations like proximity, overlap containment, adjacency, ways of measuring geographic space, and how geographical features and phenomena relate to one another. It is part of spatial literacy, a type of literacy that begins with content knowledge and encompasses an understanding of the Earth's systems, how they interact with the sphere of human influence, and how geographic space is represented. Spatial literacy allows you to use maps and other spatial tools to reason and make key decisions about spatial concepts.[4] For example, geometric visualization is a spatial literacy skill that includes calculating distances between features, calculating buffer regions (how far away one feature is from another, for example), and identifying areas or perimeters.

The Aspen Global Change Institute (*https://www.agci.org*) identifies six systems of the planet: the atmosphere, cryosphere, hydrosphere, biosphere, geosphere, and anthroposphere (or human presence on Earth). Geospatial data allows us to comprehend

---

4 To learn more about spatial literacy, see National Research Council. 2006. *Learning to Think Spatially.* Washington, DC: The National Academies Press. *https://oreil.ly/i3olt*.

the interconnectivity of all these systems, and we can use big data—lots of data—to answer well-formulated data questions.

You don't have to become an expert to retain important spatial literacy skills for bigger questions. If you understand at a fundamental level how things work in geospatial data and technology, you are already on your way to constructing more complex ideas. You will learn to formulate a data question and determine actionable steps toward developing your novel application or solution. For tools written with Python, the source code is available, and I encourage you to learn from, modify, extend, and share these and other analysis tools.

Let's look at an example. The map you see in Figure 1-3 was created as an exploration of economic precarity.

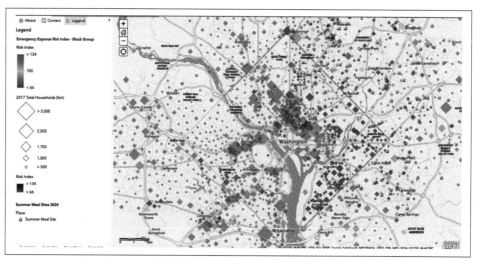

Figure 1-3. Risk Index Summer Meals (ArcGIS) targeting expansion of US Department of Agriculture summer meal programs

The red squares indicate populations in Washington, DC, where the average family could not afford a few hundred dollars to cover an unexpected emergency. These red squares show the families at the most risk for being unable to meet household expenses, and bigger squares reflect a larger number of households. The green squares are sites where, in 2020, the Summer Food Service Program (SFSP) served free meals to low-income children when school was not in session. Do you have any thoughts regarding where SFSP assigned the locations?

Looking at these layers of geospatial data together on the map in Figure 1-3, you can see the relevance of nonspatial data. You would be limited when interpreting tabular data without an understanding of population characteristics, like the size and location of the families affected and how and whether these characteristics influence the risk index.

Geographic information systems analyze data and display real-time geographic information across a wide variety of industries. Although there are similarities between spatial and nonspatial analyses, spatial statistics are developed specifically for use with geographic data. Both are associated with geographic features, but spatial statistics look specifically at *geocoded* geographical spatial data. That is, they incorporate space (including proximity, area, connectivity, and other spatial relationships) directly into their mathematics. For example, think about the kinds of data that airports generate. There are nonspatial statistics for variables such as region, use (military or civilian/public), and lists of on-time arrivals and departures. There are also spatial components, such as runway elevation and geographical coordinates.

Complex problems are spatial. Where are these problems occurring, and how can we plan for better outcomes in the future?

Map Projections

Our level of comfort in viewing maps belies their complexity. Most maps contain multiple layers of information. We can make interactive maps layering multiple datasets and experiment with how we communicate findings. But we also need to exercise caution. Maps might be familiar, but familiarity isn't the same as accuracy or competency. Projections are a great example.

Planet Earth is not perfectly spherical. That makes sense when you think of the chemical nature of the planet and how the centrifugal force caused by spinning in space tends to push out the middle, resulting in an oblate spheroid shape. Technically, the Earth's shape is an *ellipsoid*: its circumference around the poles is shorter than the circumference around the equator, almost like the planet has been squished from top to bottom. When we attempt to map the planet's surface to create a two-dimensional map, we use a geographic coordinate system (Figure 1-4) with latitude and longitude lines—called a *graticule*—to project that imperfect sphere, that coordinate system, onto a flat surface. The simplified projections account for complex factors, like how the earth's gravitational field changes with alterations in topography. We call this the *geoid*. It is important to be aware that the whole world won't fit on a piece of paper or a computer screen—at least, not in a visible, easily interpretable manner.

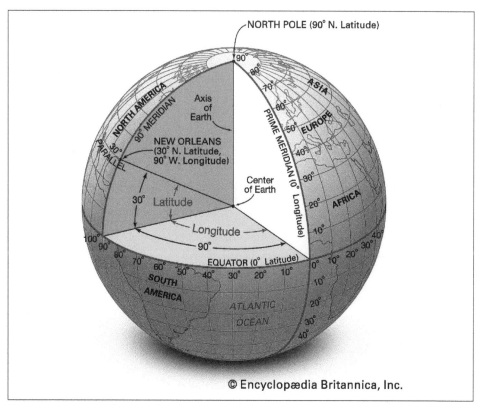

Figure 1-4. Geographic coordinate systems

These different projections include *conic*, *azimuthal*, and *cylindrical*. And if you use OpenStreetMap or Google Maps, you're familiar with the Web Mercator coordinate system. Each has advantages and disadvantages, including distortions in area, distance, direction, and size. You will be glad to know that we don't need to wrestle with these compromises alone—software manages much of the complicated math.

Getting to know the variations of these maps (Figure 1-5) will allow you to select the optimal projection for your purposes. You will always have to make compromises and trade-offs, choosing which aspect to optimize and accepting a little bit of distortion in other aspects. The popular Mercator projection (Figure 1-6) is useful for navigation but distorts the areas near the poles—which, famously, makes Greenland look enormous. You can see how different the visualizations are in the equal area projections shown in Figure 1-5, but Greenland remains at the appropriate scale in all of them. In the Mercator projection in Figure 1-6, although South America is actually eight times larger than Greenland, they appear to be similar in size.

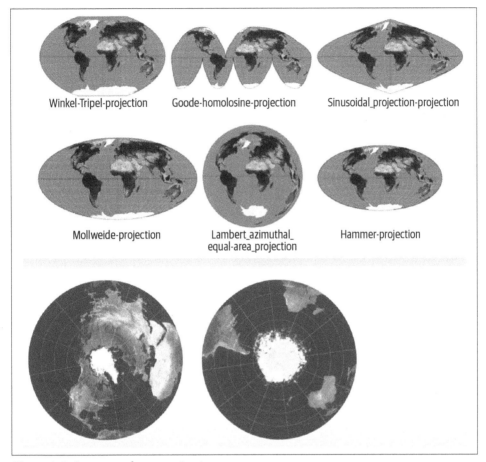

Figure 1-5. Some equal area projections

In my work in population health, area is the most critical aspect. It has to be maintained as accurately as possible on projections. When I'm mapping percentages or raw numbers on a map, I want to be as impartial as possible; if a small place looks too large compared to other places, there's an inherent bias there that affects my interpretation of that map. I'm going to do my best to look at the projection's weaknesses and its strengths and say, "I'm going to choose one that maintains area." Maps that maintain area are called *equal area projections*.

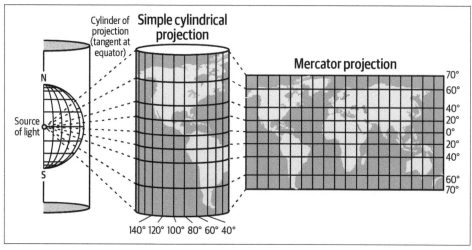

Figure 1-6. Mercator projection

If I ensure that the measures most relevant to my visualization are captured in the coordinate system I select, I'm most of the way there. Naturally, I would like my values to match the actual values in the real world as closely as possible.

Map Error Scores

A map's degree of deviation from the accuracy of a globe (which is 0.0) is called *error*. It is typically measured by the Goldberg-Gott error score,[5] which is derived from the normalized sum of the squares of the six parameters:

$$\Sigma_\in = \left(\frac{I}{N_i}\right)^2 + \left(\frac{A}{N_a}\right)^2 + \left(\frac{F}{N_f}\right)^2 + \left(\frac{S}{N_s}\right)^2 + \left(\frac{D}{N_d}\right)^2 + \left(\frac{B}{N_b}\right)^2$$

The normalization constants ($x = \lambda$, $y = \varphi$) are: $Ni = 0.51$, $Na = 0.41$, $Nf = 0.64$, $Ns = 0.60$, $Nd = 0.449$, and $Nb = 0.25$. I = isotropy, A = area, F = flexion, S = skewness, D = distances, and B = boundary cuts for some standard projections.

A lower score reflects fewer errors or compromises in the rendering of the flat map, minimizing distortion in area, direction, and distance. For reference, a globe would be 0.0. The Mercator projection (Figure 1-6) has an error score of 8.296, while the Winkel tripel projection (one of the examples shown in Figure 1-5) comes in at 4.563. The most recent J. Richard Gott map (*https://oreil.ly/wUiBK*) has the lowest error score to date: 0.881.

---

5 Gott, J. Richard, III, Goldberg, David M., and Vanderbei, Robert J. 2021. "Flat Maps that Improve on the Winkel Tripel." arXiv preprint arXiv:2102.08176.

Vector Data: Places as Objects

Before we dig deeper into how to explore vector data in Python, I need to introduce a few concepts that will be useful as we move through the book. We'll be working with *vector data,* which uses points, lines, and polygons to communicate data. We will use Python scripts and QGIS integrations to load datasets into a map and examine the structure of the vector data. I'll also show you how different tools allow you to customize maps with colors and symbols to improve clarity and accuracy.

Figure 1-7 is an ArcGIS rendering of New York City's Central Park and surrounding buildings. You can see that the geometry of each feature is represented as points, lines, or polygons. The geometry of a feature determines how it is rendered: as a point, line, or polygon. Additional information about the features might reveal the type of structure, year built, architectural dimensions, and other attributes accessible in an *attribute table.*

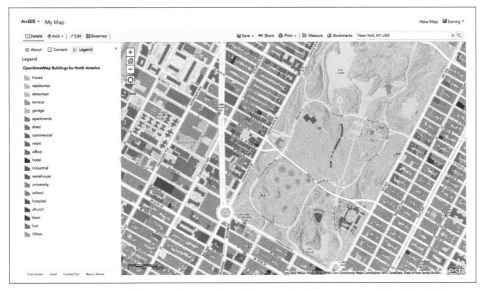

Figure 1-7. ArcGIS rendering of vector data showing building types in New York City, with attribute table

Geographical systems can work with many types of data. You may see a vector data file designated as a shapefile (formatted with the extension *.shp*) or geodatabase (*.gdb*). Lidar (light detection and ranging) surveys are collected as vector data but are often created and stored in gridded raster data formats.

Compare the file formats used with word-processing programs like Microsoft Word (*.docx*) to a simple text file (*.txt*). The content (the words on the page) might be the same in both files, but the complexity and sophistication are certainly less in the text file. If you wanted to share a piece of writing, why would you choose a text file? What if you wanted the document to be read by everyone regardless of software, or what if you wanted ease of portability or a smaller storage format?

GIS file formats work that way, too: although the content is the same, GIS file formats vary in functionality. Shapefiles do not have a topological or spatial layer, whereas with a geodatabase, such layers are optional. GIS file formats also vary in simplicity, redundancy, error detection, and storage size. Census geographic data uses TIGER/Line extracts, or shapefiles. They are grouped as a set that includes digital files (vector coordinates with an *.shp* extension), an index (*.shx*), and dBase attribute data (*.dbf*).

The most familiar coordinate projection system is longitude and latitude. These coordinates accurately describe where a particular place is on the Earth's surface. You can be dropped anywhere in the world, and if you have your longitude (X) and latitude (Y), you know your location. This precise location is called a *point feature*. A *point attribute* has X and Y values as well, but attributes can be quantitative or qualitative descriptions. The point attribute describes the feature. Optionally, a Z value can be used to represent values in three dimensions, with Z referring to elevation. When location data appears in a spreadsheet, you can use the columns for latitude and longitude to find a point. The air-quality spreadsheet shown in Figure 1-8 includes geographic data (latitude and longitude) and nongeographic data (the air-quality measurement in the value column), allowing a GIS application to add information associated with a particular geographic location.

	A	B	C	D	E	F	G	H	I	J	K
1	locationid	location	city	country	utc	local	parameter	value	unit	latitude	longitude
2	62724	Far West	Pasadena	US	2021-02-02T	2021-02-01T	pm25	27.7	¬µg/m¬≥	34.1324	-118.1834
3	62724	Far West	Pasadena	US	2021-02-02T	2021-02-01T	pm25	25.6	¬µg/m¬≥	34.1324	-118.1834
4	62724	Far West	Pasadena	US	2021-02-02T	2021-02-01T	pm25	26	¬µg/m¬≥	34.1324	-118.1834
5	62724	Far West	Pasadena	US	2021-02-02T	2021-02-01T	pm25	25.4	¬µg/m¬≥	34.1324	-118.1834
6	62724	Far West	Pasadena	US	2021-02-02T	2021-02-01T	pm25	22.8	¬µg/m¬≥	34.1324	-118.1834
7	62724	Far West	Pasadena	US	2021-02-02T	2021-02-01T	pm25	22.6	¬µg/m¬≥	34.1324	-118.1834
8	62724	Far West	Pasadena	US	2021-02-02T	2021-02-01T	pm25	22.2	¬µg/m¬≥	34.1324	-118.1834
9	62724	Far West	Pasadena	US	2021-02-02T	2021-02-01T	pm25	21.5	¬µg/m¬≥	34.1324	-118.1834
10	62724	Far West	Pasadena	US	2021-02-02T	2021-02-01T	pm25	21.5	¬µg/m¬≥	34.1324	-118.1834
11	62724	Far West	Pasadena	US	2021-02-02T	2021-02-01T	pm25	25.4	¬µg/m¬≥	34.1324	-118.1834
12	62724	Far West	Pasadena	US	2021-02-02T	2021-02-01T	pm25	23.1	¬µg/m¬≥	34.1324	-118.1834
13	62724	Far West	Pasadena	US	2021-02-02T	2021-02-01T	pm25	22.5	¬µg/m¬≥	34.1324	-118.1834
14	62724	Far West	Pasadena	US	2021-02-02T	2021-02-01T	pm25	21.2	¬µg/m¬≥	34.1324	-118.1834
15	62724	Far West	Pasadena	US	2021-02-02T	2021-02-01T	pm25	21.8	¬µg/m¬≥	34.1324	-118.1834
16	62724	Far West	Pasadena	US	2021-02-02T	2021-02-01T	pm25	26.6	¬µg/m¬≥	34.1324	-118.1834
17	62724	Far West	Pasadena	US	2021-02-02T	2021-02-01T	pm25	24.4	¬µg/m¬≥	34.1324	-118.1834
18	62724	Far West	Pasadena	US	2021-02-02T	2021-02-01T	pm25	25.2	¬µg/m¬≥	34.1324	-118.1834
19	62724	Far West	Pasadena	US	2021-02-02T	2021-02-01T	pm25	26.1	¬µg/m¬≥	34.1324	-118.1834
20	62724	Far West	Pasadena	US	2021-02-02T	2021-02-01T	pm25	25	¬µg/m¬≥	34.1324	-118.1834
21	62724	Far West	Pasadena	US	2021-02-02T	2021-02-01T	pm25	22.9	¬µg/m¬≥	34.1324	-118.1834
22	62724	Far West	Pasadena	US	2021-02-02T	2021-02-01T	pm25	23.5	¬µg/m¬≥	34.1324	-118.1834

Figure 1-8. A dataset of air-quality measurements that includes spatial and nonspatial data

Raster Data: Understanding Spatial Relationships

Vector data focuses on what is visible in a particular location. There are specific boundaries or areas on a map where data or objects are either present or absent. You wouldn't expect a building, or even a polygonal object representing a city, to exist in every location within a specific boundary.

Raster data, on the other hand, is continuous data that lacks specific boundaries but is present across the entire map view, such as imagery, surface temperatures, and digital elevations. *Raster* is data displayed as a pixelated image matrix, such as that shown in Figure 1-9, instead of the points, lines, and polygons of vector data. Each pixel corresponds to a specific geographical location. Don't worry if this seems abstract now: both types of data will be easier to visualize once we begin working with them. Perhaps slightly more intuitive, consider surface elevation, precipitation, or surface temperature. You can record these measures at each location included within your study location, regardless of your study view.

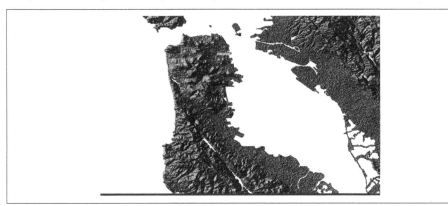

Figure 1-9. San Francisco depicted as a raster (QGIS)

So far, I am simply describing the distribution of points, polygons, or lines within our research location. Raster data is represented as an array of values divided into a grid of cells. The terms *cells* and *pixels* describe spatial resolution and are often used interchangeably. The dimension of the cell or pixel represents the area being covered. *Spatial data models* take these abstract representations of real-world objective and/or field views and explore mathematical relationships to model or predict relationships.

The big differentiator between photos and raster images is that rasters include data on the expanded bands of light wavelengths. This enhanced data, beyond simply red, green, and blue wavelengths, allows machine-learning models to distinguish among a wide variety of objects. This is because different objects reflect infrared light in different ways, yielding additional information in a multispectral image. Many space agencies around the world make data from their Earth observation satellites available freely. These datasets are immensely valuable to scientists, researchers, governments, and businesses.

A *hillshade* raster, as seen in Figure 1-9, uses light and shadow to create the 3D effect of the area in view.

It is necessary to consider multiple concepts simultaneously when reviewing a systems-level approach to GIS. The smaller components working within a larger system interact dynamically to reveal patterns in the system. Geometric visualization, for example, includes calculating distances between features, calculating buffer regions (how far away one feature is from another, for example), and identifying areas or perimeters. It will simplify our discussion if you think of these topics as parts of a whole. Understanding these introductory concepts will simplify your learning in the following chapters. It is important to begin with a baseline of spatial literacy.

Evaluating and Selecting Datasets

There are many datasets to explore for use in tutorials for learning a new skill, following along in a new application, or even launching your own independent geospatial project. The datasets in this book have been vetted and found workable on a wide variety of applications and workflows.

Before selecting your dataset, you need to evaluate your options. Information about a dataset is called *metadata*. Often there is also a supplemental data file that describes attributes like field headings. This is called a *data dictionary*. You can explore an example: the Landsat Data Dictionary (*https://oreil.ly/BgBaG*), published by the US Geological Survey (USGS).

You can learn a lot by looking through metadata. You might think of metadata as the label on a can of soup: you read it because you want to know what the ingredients are and whether the soup is good for you. Figure 1-10 shows an example of metadata. The most important information you'll want to check should include the geographic area shown, the attributes listed, the map projection the dataset uses, its scale, and whether there is a fee to use it.

I suggest first attempting to follow along with the suggested data resources listed in this book. Once you feel confident, explore datasets related to your interests and see what you can discover.

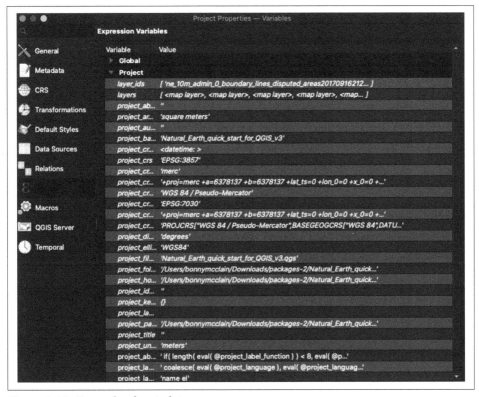

Figure 1-10. Example of metadata

Summary

The field of geospatial analysis is vast, and Python is an expansive topic about which much has been written. It's hard to imagine any single book introducing either of these topics, let alone both, in a complete and authoritative manner. So I won't attempt to do so in this book; instead, the goal here is to explain important foundational elements and introduce you to open source tools and datasets that you can use to answer geospatial questions.

This chapter presented an overview of important geospatial concepts like coordinate systems, projections, and the two main types of geospatial data: vectors and rasters. You also began to learn how to think spatially, and we finished with an introduction to datasets and how to choose what data to work with. Don't be alarmed if this all seems like a lot! Right now, I just want to launch your curiosity by sharing the possibilities of working with open source geospatial data using Python.

## CHAPTER 2

Essential Facilities for Spatial Analysis

Chapter 1 covered the complexity of mapping a three-dimensional (3D) globe to a two-dimensional (2D) coordinate system. This often requires an understanding of how to select an appropriate map projection depending on the area you are interested in viewing or analyzing. You learned how 3D coordinates on the surface of the Earth can be converted to 2D coordinates. For that, the concepts of *geoid* and *ellipsoid* were introduced. You also saw a variety of map projections in Chapter 1, and I discussed how to choose an appropriate one for a given area—for example, selecting a projection that minimizes distortion over the area you are viewing.

What happens to topographical features that are below or above the Earth's surface? Geospatial tools are also able to analyze these features spatially as well as time-series data, which is often described as a *fourth dimension.*

You may have noticed the term *GIS* doing a lot of the heavy lifting when learning about spatial literacy, but in reality there are multiple integrated concepts contributing at the systems level. I have discussed the *spatial data framework,* where spatial locations are identified on the Earth's surface. This is fundamental for building a comprehensive (and reliable) reference system for geocoding and mapping data.

In this chapter, you will continue to build on your spatial literacy skills with three brief sample projects in QGIS. First, you will download and customize the QGIS dashboard. Then I'll introduce simple Python scripting as you explore data from the NYC Open Data portal. The Mapping Inequality project will provide an opportunity to explore historical data and download the data for analysis in QGIS. This will help you familiarize yourself with QGIS and introduce you to important concepts, including uploading vector data layers, filtering datasets, and writing Python scripts in the console.

Exploring Spatial Data in QGIS

I mentioned QGIS in the previous chapter, and although you will work with it extensively in Chapter 3, you'll get your feet wet in this chapter by using QGIS and Python to explore publicly available datasets and discuss some key concepts of spatial literacy.

QGIS is a free and open source GIS. A *GIS*, which you might recall stands for "geographic information system," is a complete application or system of tools for working with geospatial data. Once you download the software, you will have access to these tools. Python is available as a plug-in and is easily installed.

Installing QGIS

To follow along, please install QGIS. You can download and install the software at the QGIS Project's website (*https://qgis.org/en/site*). (Look for the long-term repositories for stability, identified as "LTR" on the site. Detailed installation instructions are in the User Guide (*https://oreil.ly/mgycA*).) Your dashboard will resemble the one shown in Figure 2-1 once you add a few panels to the view.

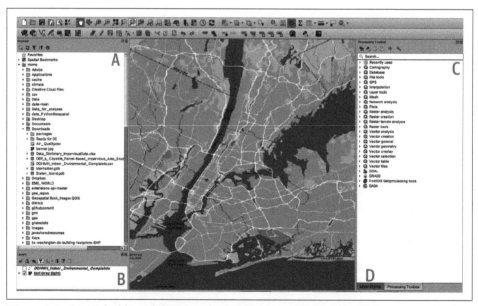

Figure 2-1. QGIS dashboard showing the (A) Browser, (B) Layers, (C) Processing Toolbox, and (D) Layer Styling panels

Figure 2-1 shows the (a) Browser, (b) Layers, (c) Processing Toolbox, and (d) Layer Styling panels (unselected in the lower right corner). The Browser allows you to view your computer file system and upload data into the QGIS Layers panel or drag files directly onto the canvas. The Processing Toolbox has several tools to explore; you will be using the Python scripting tool in a later exercise in the chapter. Last, the Layer Styling panel provides customization for points on a map, transparency of levels, and color schemes (among its many features). In the last exercise in the chapter, you will assign a color scheme to a map legend describing grades of perceived desirability of neighborhood communities.

Customizing your dashboard is relatively straightforward. Select View from the menu at the top of your screen and scroll down to Panels. Figure 2-2 shows the checked options that have been added to the console. You can move them around according to your preference by selecting them individually and moving them to the desired location.

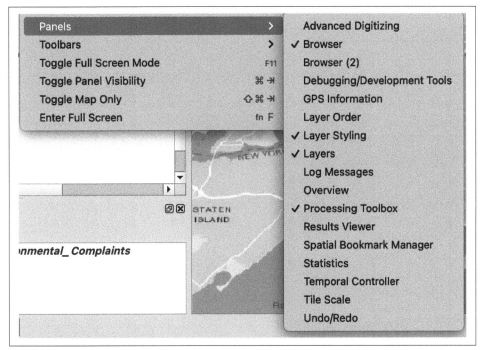

Figure 2-2. Customizing your dashboard with display options, such as panels

Adding Basemaps to QGIS

Adding additional basemaps is one last customization that will help enhance your work environment. Basemaps are reference maps that add context to the data layers. Your XYZ Tiles folder (scroll down in your lefthand Browser panel) will currently appear empty. To add basemaps, navigate to the GitHub page of Qiusheng Wu (*https://oreil.ly/SH2p6*), Assistant Professor of Geography at the University of Tennessee, Knoxville. Click on the raw file to download it directly to the Downloads folder.

You can opt for a different location, but you will have to recall where it is located to upload it to the Python console. Figure 2-3 is a snippet of the screen you will see.

```
"""
This script should be run from the Python consol inside QGIS.

It adds online sources to the QGIS Browser.
Each source should contain a list with the folowing items (string type):
[sourcetype, title, authconfig, password, referer, url, username, zmax, zmin]

You can add or remove sources from the sources section of the code.

Script by Klas Karlsson
Sources from https://qms.nextgis.com/

Licence GPL-3
"""

from qgis.PyQt.QtCore import QSettings
from qgis.utils import import iface
```

Figure 2-3. The executable file for downloading basemaps to QGIS

Right-click in the text and save the file to your Downloads folder as a *.py* script (Figure 2-4).

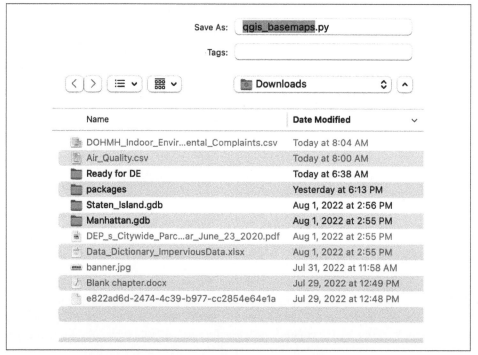

Figure 2-4. Saving the Python script to your Downloads folder

Your script is now saved to your local computer. There are multiple locations to open the Python Console in QGIS, indicated by the Python icon. Figure 2-5 highlights the console icon in the Plugins menu at the top of the dashboard.

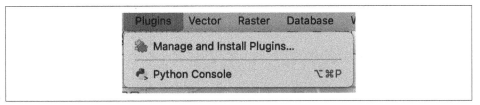

Figure 2-5. Installing the Python plug-in to QGIS

The Python Console is where you can enter a few line-by-line commands or click the Show Editor button on the menu to enter Python scripts or snippets of code. The code snippets are run from the editor. The Python code editor has a save option if you are writing longer scripts. This feature is a reason you may decide to run code in the editor as opposed to simply dragging layers onto canvas.

Click the icon of a pad of paper with a pencil to open the scripting window. When you select the green + symbol, you can navigate to the location of the downloaded script. Hit the run icon (a little green arrow symbol), and the code will run in the console (Figure 2-6). You will now see available basemaps populating the XYZ Tile folder. Drag them directly onto the canvas to view them. Be sure to reorder the basemap when adding layers to the canvas. You will want the data layers to be at the top of the Layers panel.

Figure 2-6. Run the script and the basemaps will populate the XYZ Tiles folder

QGIS has a powerful programming interface that allows you to extend its core functionality and write scripts to automate your tasks in Python scripting language. Even if you are new to programming, learning a little bit of Python and navigating this interface will enable you to be much more productive in your geospatial data work.

Exploring Data Resources

This book mainly uses data resources that I first used in my training or in projects, updated with the latest reported data. You can access the resources directly from my GitHub page (*https://oreil.ly/SbS0R*). I want to be certain you can access the datasets no matter what operating system you use. QGIS has a wide variety of tools that make this straightforward.

If you are an advanced user and would like to replace the datasets in this book with local data or data that reflects your personal or professional interests, here are a few suggestions:

- You will want a healthy mix of urban and rural locations with different types of roads, boundaries, water features, and topographies.

- When given a choice, I prefer recent data, but for instructional purposes, I defer to sources with the widest variety of attributes.

- Don't forget to save your data in the GeoPackage file format so you will be able to continue exploring.

- If you are creating a raster file of your region, locate the extent coordinates of your largest layer.

Visualizing Environmental Complaints in New York City

You can see an example of how geography is represented in QGIS in Figure 2-7. The points represent indoor environmental complaints received by the Department of Health and Mental Hygiene (DOHMH) in New York City. The data in this image is from NYC Open Data (*https://oreil.ly/QGsn5*), a resource for free public data from city agencies and other partners. This particular dataset can be downloaded from the GitHub repository accompanying this book (*https://oreil.ly/9ADWy*).

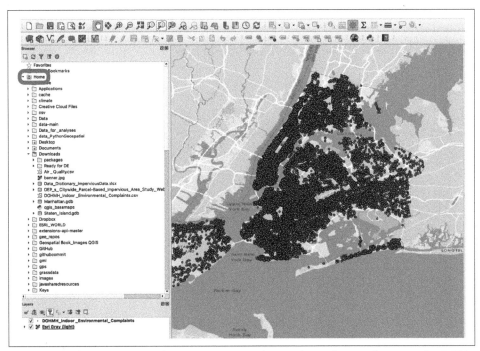

Figure 2-7. The DOHMH Indoor Environmental Complaint dataset shown in the QGIS interface; the left panel is the Browser, and the right panel is the canvas

Uploading Data to QGIS

You have a few options to upload the data into QGIS. Using the Browser panel, select the Home folder (circled in Figure 2-7), navigate to your Downloads folder, and select the dataset file that you downloaded. You can drag shapefiles (*.shp*) into the view from the Browser directly onto the canvas (the right panel).

What you are observing in Figure 2-7 are the 311 information hotline calls made within New York City during a certain period of time. Initially, you don't know much about the nature of the calls or how the data could be filtered for better insights.

Uploading files with the Data Source Manager

The Data Source Manager (circled in Figure 2-8) is the first option, and my preferred one, for uploading data from *.csv* files. If your data has a geometry column, as this one does (see the Latitude and Longitude columns in the Sample Data window in Figure 2-8), you can select them in the X field (longitude) and Y field (latitude) and add the data to the Layers panel in your console.

Until you confirm or edit the delimited file with geometric attributes, it will not be considered spatial. This option is preferred for *.csv* files because, although QGIS often recognizes geometric attributes instinctually, you may need to change data types manually in the column headings.

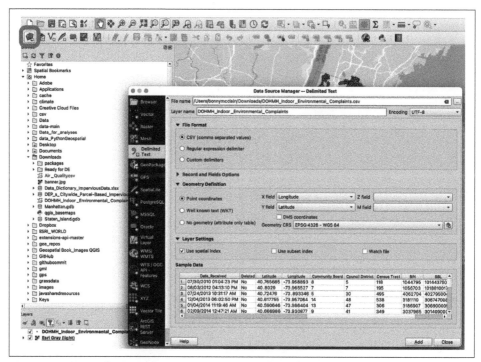

Figure 2-8. Using the Data Source Manager to upload delimited text files

Adding data as a vector layer

You also have the option of adding the dataset as a vector layer, as shown in Figure 2-9. This works well for GeoJSON and a variety of other data formats. QGIS can detect most types of geometries. I prefer this method when working with PostgreSQL databases, as it simplifies the process of updating files stored outside of QGIS.

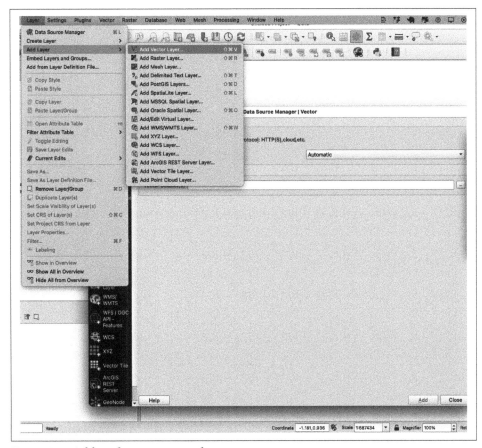

Figure 2-9. Adding data as a vector layer

Right-click on a data file in the Layers panel and you will see an option to view the Attribute Table (shown as a snapshot in Figure 2-10). The Attribute Table is useful for reviewing the columns available as well as the format of the geometry (Latitude and Longitude). A small but important detail is to match the case of column headers when writing code to avoid errors when running your scripts.

	Complaint_Number	Address_Street	d_Address_Street	Incident_Address	cident_Address_Z	ient_Address_Bor	omplaint_Type_31	Descriptor_1_311	Complaint_Status	Date_Received	Deleted	Latitude	Longitude	Com
34	T-12-06-25-...	126	EAST 54 STR...	126 EAST 54 ...	10022	Manhattan	Asbestos	N/A	05-Close	06/25/2012 0...	No	40.759148	-73.97091	5
35	T-12-05-29-...	128	EAST 56 STR...	128 EAST 56 ...	10022	Manhattan	Asbestos	N/A	05-Close	05/29/2012 1...	No	40.760441	-73.970093	5
36	T-18-02-23-...	136	EAST 56 STR...	136 EAST 56 ...	10022	Manhattan	Asbestos	N/A	05-Close	02/23/2018 0...	No	40.760342	-73.969859	5
37	T-19-09-07-...	301	PARK AVENUE	301 PARK AV...	10022	Manhattan	Asbestos	N/A	05-Close	09/07/2019 1...	No	40.756648	-73.974358	5
38	T-10-09-07-...	130	BARUCH PLA...	130 BARUCH...	10002	Manhattan	Asbestos	N/A	05-Close	09/07/2010 0...	No	40.718299	-73.976775	3
39	T-16-08-23-...	90	BARUCH DRI...	90 BARUCH ...	10002	Manhattan	Asbestos	N/A	05-Close	08/23/2016 1...	No	40.718211	-73.97772	3
40	T-21-12-02-...	75	BARUCH DRI...	75 BARUCH ...	10002	Manhattan	Asbestos	N/A	05-Close	12/02/2021 0...	No	40.717424	-73.978276	3
41	T-17-08-24-...	250	WEST 34 STR...	250 WEST 34...	10119	Manhattan	Asbestos	N/A	05-Close	08/24/2017 0...	No	40.751771	-73.992489	5
42	T-22-01-23-...	401	7 AVENUE	401 7 AVENUE	10001	Manhattan	Asbestos	N/A	05-Close	01/23/2022 0...	No	40.749794	-73.991486	5
43	T-12-10-15-0...	455	MADISON AV...	455 MADISO...	10022	Manhattan	Asbestos	N/A	05-Close	10/15/2012 11...	No	40.758114	-73.975393	5
44	T-12-02-06-...	611	6 AVENUE	611 6 AVENUE	10022	Manhattan	Asbestos	N/A	05-Close	02/06/2018 0...	No	40.768002	-73.977577	5
45	T-20-06-02-...	685	FIFTH AVENUE	685 FIFTH A...	10022	Manhattan	Asbestos	N/A	05-Close	06/02/2020 ...	No	40.76082	-73.975093	5
46	T-16-04-01-...	425	WEST 33 STR...	425 WEST 33...	10001	Manhattan	Asbestos	N/A	05-Close	04/01/2016 1...	No	40.753091	-73.99764	4
47	T-18-12-05-...	320	WEST 31 STR...	320 WEST 31...	10001	Manhattan	Asbestos	N/A	05-Close	12/05/2018 1...	No	40.750637	-73.99568	4
48	T-21-11-09-0...	440	WEST 34 ST.	440 WEST 3...	10001	Manhattan	Asbestos	N/A	05-Close	11/09/2021 0...	No	40.754022	-73.997842	4
49	T-17-09-09-...	30	ROCKEFELLE...	30 ROCKEFE...	10112	Manhattan	Asbestos	N/A	05-Close	09/09/2017 0...	No	40.758754	-73.978692	5

Figure 2-10. Examining the Attribute Table to determine geometry columns

You can also customize your basemaps now that you have uploaded the additional basemaps to QGIS (as shown in Figure 2-11).

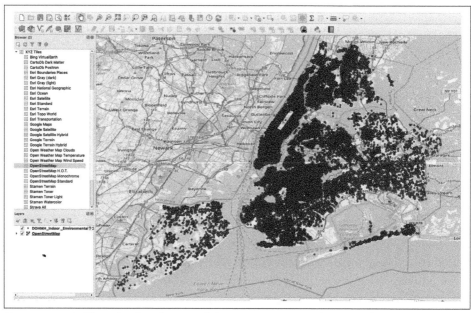

Figure 2-11. Adding the data layer and the OpenStreetMap basemap from XYZ Tiles

Setting the Project CRS

QGIS will attempt to assign a project coordinate reference system (CRS), as seen in Figure 2-12. You can change it in the Layer CRS window, visible when clicking the data in the Layers panel.

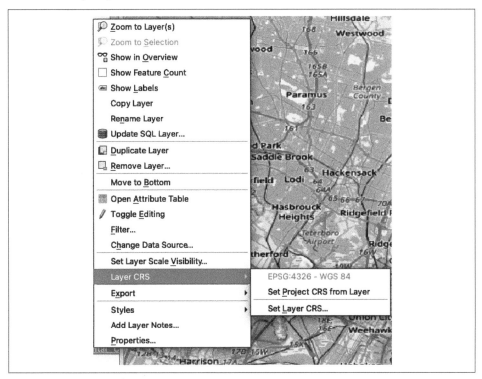

Figure 2-12. Setting the CRS Layer EPSG:4326—WGS84

Now that the CRS layer is selected, you can hover over points in the map for additional information. Select the Identify Features icon (the small "i" inside a blue circle), then select a point on the map.

You will explore many of these geospatial and nongeospatial features a little later on when we begin working in the Python Console.

Using the Query Editor to Filter Data

Let's say that our research question here is: *where are Manhattan residents reporting issues concerning asbestos?* Asbestos levels are important to monitor because asbestos is a known carcinogen when released into the atmosphere. QGIS has a built-in query builder that can help familiarize you with datasets. Right-click on the dataset in the Layers panel and choose Filter.

When you click on a field, it will populate in the Provider Specific Filter Expression console, shown in Figure 2-13. Select the Sample option under Values to view the sample data. Add an Operator and click the Test button at the bottom of the console. You will see how many rows have been returned. You can add more filters to the data or click OK and load the updated dataset to your canvas. Scrolling through the Fields and selecting Sample will provide information for exploring in the console.

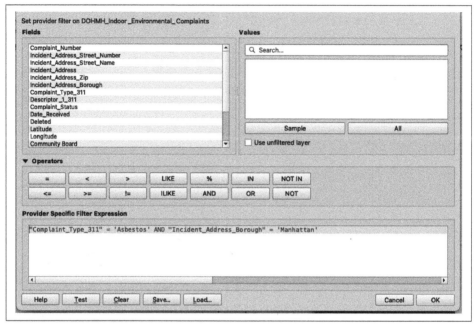

Figure 2-13. Filtering data by incident type and location

The canvas in Figure 2-14 has been filtered to show only asbestos complaints. There are additional options to filter based on Complaint_Status and Date, if you want to explore!

What follows is a brief introduction to the Python Console in QGIS. You will run some code and even generate a template script in the Processing Toolbox.

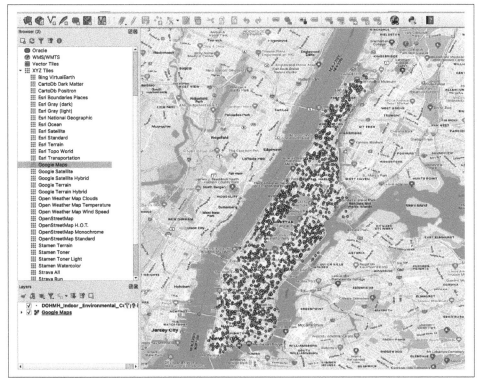

Figure 2-14. The canvas updated with a new query to show asbestos complaints

Visualizing Population Data

Whether you are planning on making maps professionally or for personal use, finding great data resources can be a challenge. Natural Earth (*https://www.naturalearth data.com*) is a public-domain dataset available through collaboration and support from volunteers and the North American Cartographic Information Society (NACIS).

Its freely available vector and raster data—in shapefile and TIFF format, respectively—all use the EPSG:4326 projection, also known as the WGS84 projection. Additionally, you can choose how detailed the data is for downloading. The more detailed data, 1:10 m,[1] will be the most resource intensive, with 1:50 m and 1:110 m providing moderate and coarse levels of detail. This relationship is described as a representative fraction scale: 1 unit on the map (such as inches or centimeters)

---

1 The scale of a map is the ratio of a distance on the map to the corresponding distance on the ground. For reference, *1:10 m* means that 1 unit on your map (screen) is equal to 10,000,000 units. 1 m on the map = 10,000,000 m. Read more about Map Scales (*https://oreil.ly/LJFbr*) at the US Geological Survey Information Center.

represents the equivalent units on the ground. In our example, 1:50,000,000 means 1 cm = 50 km or 1 inch = 790 miles. In the left panel in Figure 2-15, you see the Browser window. Once you download the dataset, it can be uploaded from the Browser window and brought directly onto the canvas (in the right panel). Drag the shapefile (.*shp*) into view. The points represent populated places in the world. The attributes associated with each symbol include capitals, major cities, and towns, for example. The shaded polygons represent urban areas.

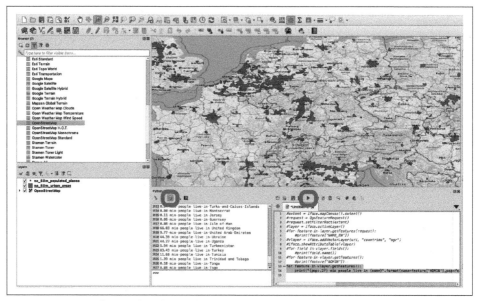

Figure 2-15. Population data by Python Console window (QGIS)

Once the data is on the canvas, you can begin to query in the Python Console. To expand the console, select the icon on the left (under the red arrow on the left). Code will run when you select the green button (under the red arrow on the right). When running a single line of code, the >>> console is fine. When I am iterating over multiple lines of code, I use the editor (the panel on the lower right).

The QGIS Python Console

You can also upload data to the canvas using the Python Console. Add the vector layer to your local computer: select Layer >> Add Layer >> Add Vector Layer. Copy the path from there or, if you know your directory and the name of the file, you can enter it directly.

Create a variable called uri (or whatever you prefer; I like uri, which stands for *uniform resource identifier*—a character string for identifying resources). You are storing the string that describes the path or source of the data. This will be one of the three parameters you see below in the code snippet for the `iface.addVectorLayer` function. The second parameter is the layer name you choose, and `ogr` is the provider key (representing the OGR Simple Features Library, today known only by its acronym). Other examples of provider keys include *postgres* and *delimited text*.

Write this code into the Python Console, substituting your own username as it appears in your filepath:

```
uri="Users/yourfilepathusername/Downloads/50m_cultural/ne_50m_populated_places.shp"
vlayer = iface.addVectorLayer(uri, "places", "ogr")
```

Run the script, and the canvas populates with the data points. You can also query populations in the editor:

```
for feature in vlayer.getFeatures():
    print("{pop:.2f} mio people live in {name}".format(name=feature['ADMIN'],
        pop=feature['POP_EST']/1000000))
```

The indentations in your code are important for delineating code that is part of a function; you will get an error if you forget to include them when required. You can see code examples in the editor in Figure 2-15, in the Python documentation (*https://oreil.ly/aPqZE*), and in the PyQGIS Developer Cookbook (*https://oreil.ly/XFgS7*) online reference. The code result displays in the left window. When you successfully run the code, you will see the output available for scrolling.

When writing code in the editor, comment out any code you do not want to run again (by starting the line with #), as each time you run the cell it will generate output. Remember to use your own filepath. If the points are not styled to your taste, you can use the Layer Styling panel or write Python code into the console.

A renderer modifies how the data is displayed in a data layer. Next, run the following code and adjust the `setSize` parameter to your liking:

```
vlayer.renderer().symbol().setSize(6)
vlayer.triggerRepaint()
```

The `triggerRepaint` function will update the map. If you don't call this function, the map won't update until you run the canvas again.

The Class `QgsSimpleMarkerSymbolLayerBase` (*https://oreil.ly/gxOJ9*) modifies the shape, size, angle, and scale method of parameters, with many shape options available (listed in the documentation (*https://oreil.ly/5zu3W*)). Enter the following code to change the shape of the point from a circle to a star:

```
vlayer.renderer().symbol().symbolLayer(0).setShape
(QgsSimpleMarkerSymbolLayerBase.Star)
vlayer.triggerRepaint()
```

More advanced scripting options are available in the Processing Toolbox. If you scroll down to the Python Scripts icon, you will see any saved scripts listed under Example Scripts. If you select the Python icon at the top of the Processing Toolbox, you will see four options: Create New Script, Create New Script From Template, Open Existing Script, and Add Script to Toolbox.

Select Create New Script From Template. Run the script (see the log in Figure 2-16), and the output layer will show in the Layers Panel as a new layer.

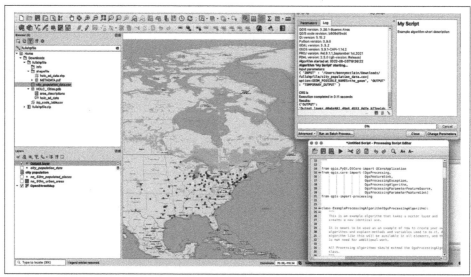

Figure 2-16. Python scripting window rendering a map

Loading a Raster Layer

The data you are using is from the QGIS sample data repository (*https://oreil.ly/3fZQh*). The steps for loading a raster layer are similar to those for loading a vector layer. The biggest differences are that we are now using a *.tif* file, and the provider parameter is now `gdal` (from the Geospatial Data Abstraction Library, which we'll discuss in Chapter 9) instead of `ogr`:

```
uri="Users/bonnymcclain/Downloads/qgis_sample_data/raster/SR_50M_alaska_nad.tif"
rlayer = iface.addRasterLayer(uri, "SR raster","gdal")
```

It can be helpful to use `if` statements in determining if the code run is successful. The `print` statement is indented and will be run when the specific condition holds:

```
if rlayer.isValid():
        print("This is a valid raster layer!")
else:
        print("This raster layer is invalid!")
```

Following is an example of raster statistics you can run in the console, and many more are available in the QGIS documentation (*https://oreil.ly/OKP96*):

```
print("Width: {}px".format(rlayer.width())
print("Height: {}px".format(rlayer.height()))
print("Extent: {}".format(rlayer.extent().toString()))
```

The output is shown in Figure 2-17 and here:

```
Width: 1754px
Height: 1394px
Extent: -6232946.6726976688951254,-735684.6617672089487314 :
6363148.4376372583210468,9275122.9686814155429602
```

This indicates the pixel values and the *extent,* or actual dimensions, of the layer we are viewing.

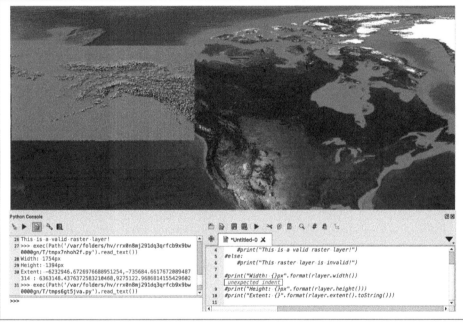

Figure 2-17. Adding a raster layer to the canvas

Redlining: Mapping Inequalities

I'll end this chapter with a powerful visualization that demonstrates the value of place and location when considering a data question. The Mapping Inequality: Redlining in New Deal America (*https://oreil.ly/LOcDk*) research project presents data from the Home Owners' Loan Corporation (HOLC) (*https://oreil.ly/M4kIq*), a federal foreclosure-prevention program introduced in 1933 as part of President Franklin Delano Roosevelt's New Deal. Here's how the site introduces the data:

> HOLC staff members, using data and evaluations organized by local real estate professionals—lenders, developers, and real estate appraisers—in each city, assigned grades to residential neighborhoods that reflected their "mortgage security" that would then be visualized on color-coded maps. Neighborhoods receiving the highest grade of "A"—colored green on the maps—were deemed minimal risks for banks and other mortgage lenders when they were determining who should receive loans and which areas in the city were safe investments. Those receiving the lowest grade of "D," colored red, were considered "hazardous."

This practice, called redlining, locked existing land-use systems into place. Areas outside of the best classifications (green or blue) were more likely to flood (due to more asphalt covering the ground), to have fewer parks and limited treelines, and to be closer to garbage dumps and interstate freeways.

The project's website allows you to select a city and explore maps. I spend a lot of time in New York City, so I selected Manhattan in the screenshot in Figure 2-18. I have walked the neighborhoods of New York City for many years, watching economic investments into neighborhoods once widely considered too dangerous to roam. The 1930s redlining maps highlighted the neighborhoods that surround Central Park as "still desirable" and "best"—as long as they weren't too close to Harlem and other primarily Black neighborhoods.[2] Although mortgage-security redlining is no longer a government-sponsored practice, the land-use systems in many of these areas are still locked in and are often explored as foundational to research directed toward environmental racism.

---

2 To learn more, see Mitchell, Bruce, and Franco, Juan. 2018. *HOLC 'Redlining' Maps: The Persistent Structure of Segregation and Economic Inequality.* National Community Reinvestment Coalition. *https://ncrc.org/holc*.

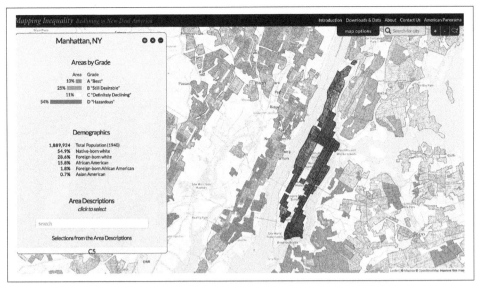

Figure 2-18. Redlining: a map of HOLC neighborhood grades in 1930s Manhattan, generated with the Mapping Inequality project's online tool

The data can be downloaded and uploaded to QGIS. Add the shapefiles by dragging them from the Browser panel onto the canvas. To update the colored polygons according to HOLC grade, use the Layer Styling panel, as shown in Figure 2-19, to select Categorized as a format and holc_grade as the value, then each color as the symbol (selected individually for each value by clicking on each square).

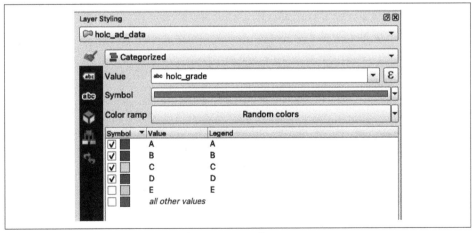

Figure 2-19. Styling the shapefiles by HOLC grade

Figure 2-20 shows the New York City map and attribute table displayed in QGIS. Saving the layer as a file will make it available for other maps, where you can examine the persistent impact of redlining.

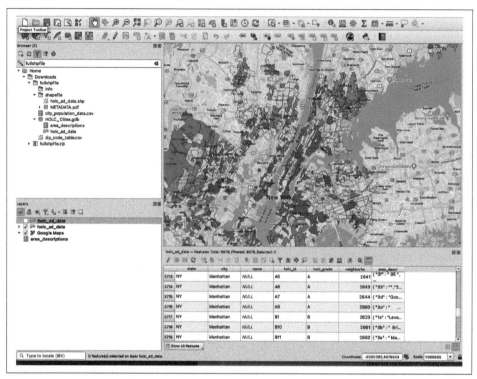

Figure 2-20. QGIS attribute table for map, which includes the attributes state, city, holc_grade, neighborhood, *and* area

Redlined areas were—and are—also more likely to be *urban heat islands*: artificial or natural locations with higher temperatures than the city average. Neighborhoods with more greenery, fewer impervious surfaces covering the ground, or fewer industrial or large buildings tend to absorb sun more easily than neighborhoods full of concrete, so they are cooler in the summer. Cooler neighborhoods are shown in yellow in Figure 2-21, and the warmer areas are indicated by orange and red. The red circle in Figure 2-21 depicts a concentration of higher surface temperatures that maps to HOLC red and yellow grades from the 1930s. The satellite images often capture urban clusters on Earth examining the LandScan urban extant and associated temperatures.

Figure 2-21. Urban heat islands, calculated with mean average between 2003 and 2018. Data source: Earth Engine Data Catalog (https://oreil.ly/Fw1tg)

Summary

In this chapter, you were introduced to QGIS and its Python Console. You learned how to load *.csv* and delimited files, shapefiles, and raster data. QGIS has a robust query editor, but often you can execute the same tasks in the Python Console or editor. You also got a taste for how bringing data together into a map can help you explore complex research questions. Publicly available geospatial datasets can be powerful decision-resource tools for exploring societal harms and benefits as well as frameworks for evaluating potential risks in the public and private sector. This book will provide a window into the scale of the resources available and show you how to interact and explore location intelligence to draw out deeper insights and bigger questions.

In the next chapter, you will learn more about PyQGIS, the Python language of QGIS.

## CHAPTER 3

QGIS: Exploring PyQGIS and Native Algorithms for Spatial Analytics

In this chapter, we will continue to focus on QGIS as you level up your skills with Python and QGIS. These are foundational concepts, and learning them will be critical to future geospatial analysis skills, including spatial algorithms, data engineering, prediction modeling and forecasting, and machine learning.

The QGIS integration of Python is called *PyQGIS*, a Python API that uses defined protocols and customization to automate workflows. Automation is important when running large scripts or building applications. The integration with an API allows you access to a large variety of datasets for exploration and analysis. You can create, modify, and query digital objects of interest that represent features in the real world.

PyQGIS is a wrapper around the underlying C++ library. All the methods and class variables implemented by the C++ version of `QgisInterface` are made available through the Python wrapper. PyQGIS *classes* target functionality within QGIS. You do not need a separate installation of Python because it is installed with QGIS directly into your system.

Python is an object-oriented programming language. You can think of objects as chunks of data (*attributes*) and behaviors (*methods*). In Python, objects also include functions. Attributes can be either data or functions. Methods are also attributes, which means you can store them in variables, just like any other attribute. A Python class describes instructions for how to change the state of an object and the attributes of the object.

I will revisit these concepts when we briefly explore Python scripting templates later in the chapter, but I will also highlight them when we use methods or functions that are defined by the QgisInterface class. We will begin by using PyQGIS to navigate a sample project. You will upload data layers and learn how to interact with them using the Python Console.

Exploring the QGIS Workspace: Tree Cover and Inequality in San Francisco

You learned about urban heat islands in Chapter 2, and we're going to expand on that here. It is well known that neighborhoods with less tree cover tend to be hotter, often leading to increased health risks. We're going to explore that idea in the first example of this chapter as it plays out in one US city, San Francisco, with the research question: *which neighborhoods in San Francisco are less likely to have tree cover?*

The map in Figure 3-1 has four layers of data superimposed onto it: one delineating neighborhood boundaries, one that provides data about tree cover in San Francisco, one that provides data about income level and race, and an OpenStreetMap for location context. The purple lines indicate neighborhood boundaries represented as a summary feature class.

One last thing we need is a *proxy*, or stand-in, for low-income neighborhoods. Is there a measure already in use that will tell us what we need to know? In fact, there is: Equity Strategy Neighborhoods (*https://oreil.ly/HeLPB*). The San Francisco Municipal Transportation Agency (SFMTA) uses this measure in applying equity policies that attempt to address disparities in transit performance. These neighborhoods were identified based on their percentage of low-income households and public housing and residents' access to personal vehicles, race/ethnicity, and disability. This could serve our purpose as a proxy for low-income neighborhoods.

With these layers in place, we can think about where areas of adequate tree cover are located and compare them to neighborhoods of different income levels and racial composition.

Figure 3-1. Urban heat islands in San Francisco, generated in ArcGIS Online

You'll be looking at sample data from DataSF (*https://oreil.ly/pLNTb*) (pictured in Figure 3-2). Download the shapefiles from these three data resource links: Equity Strategy Neighborhoods (*https://oreil.ly/HPve3*), SF Urban Tree Canopy (*https://oreil.ly/RIKUa*), and SF_neighborhoods (*https://oreil.ly/qNEsQ*). You will return to these files as soon as you set up your workspace.

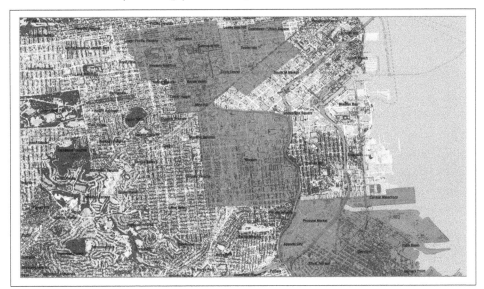

Figure 3-2. Sample data visible in the QGIS map canvas

The Python Plug-in

Open the Python Console, just as you did in Chapter 2, by either selecting the icon in the toolbar at the top of your window or opening Plugins >> Python Console from the menu bar. You can click the console, and the plug-in will be added to your work space, like in Figure 3-3.

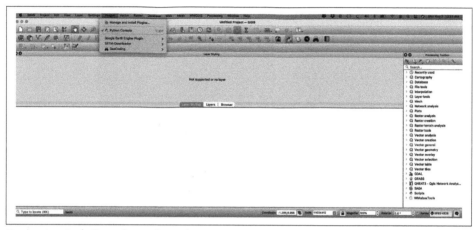

Figure 3-3. Installing the Python plug-in

The QgisInterface class provides methods for interacting with the QGIS environment. When QGIS is running, a variable called iface is set up to provide an object of the QgisInterface class to interact with the running QGIS environment. This interface allows access to the map canvas, menus, toolbars, and other parts of the QGIS application. Both the Python Console and plug-in can use iface to access various parts of the QGIS interface.

In the QGIS desktop application, iface.activeLayer() gives access to the currently selected layer in the legend. The most common use of the iface class is to get a reference to the canvas where maps are displayed.

The prompt at the bottom left of Figure 3-4 is where you will enter your short code snippets. The results of your queries will appear in the upper console. The code editor, on the right, accommodates longer lines of code and lets you work with your code a bit before you run it.

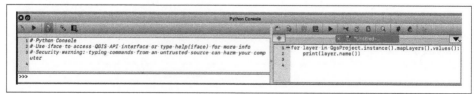

Figure 3-4. Python Console (left) and a Python code editor (right)

Reading the toolbar icons from left to right in Figure 3-4, you will see Clear Console (removes the contents in the console), Run Command, Show Editor, Options, and Help.

Selecting Options in the toolbar lets you set the font and perform additional customization. The code editor also lists icons for functions, which you can use when creating scripts directly in the editor window. A full list is given in Table 3-1.

Table 3-1. Function icons in QGIS

	Clear Console
	Run Command/Run Script
	Show Editor/Open in External Editor
	Options
	Python Console Help
	Open Script
	Save
	Save As
	Cut
	Copy
	Paste
	Find Text
#	Comment
	Uncomment
	Object Inspector

Type into the console at the >>> prompt:

```
print("Save the Planet!")
```

Hit Enter and you get the output: Save the Planet!

I also want to point out that the string "Save the Planet!" appears in red in the GQIS console when you type in this code, while the rest appears in black. This is called *syntax highlighting*, and it's a helpful tool that makes it easier to spot if you've made any typing mistakes in your code. Brackets will also be highlighted, because it is a common mistake to accidentally leave out a required opening or closing bracket.

Accessing the Data

Before we explore a multilayer example, let's load a vector layer using Python. This is the neighborhood layer from Figure 3-1. You will need to know the URL from where you downloaded your files. If you go to the Browser panel and locate the downloaded file, the URL is retrievable from Layer Properties, as shown in Figure 3-5.

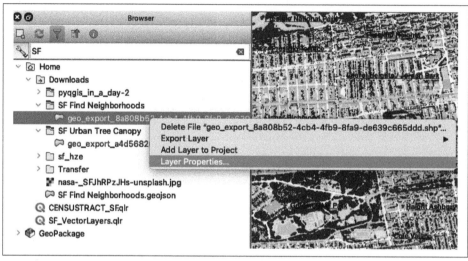

Figure 3-5. Loading data from the Browser

You'll have to tell `iface` to add the vector layer. In the code snippet that follows, `ogr` is the provider key name you saw in Chapter 2. You will work with a few different provider keys in this chapter. Replace *Path to your shape file.shp* with the URL you retrieved from the Layer Properties. You'll also create a variable for San Francisco, SF, and store the location information in it, which lets you refer to it later without having to reenter the string:

```
for layer in QgsProject.instance().mapLayers().values():
    print(layer.name())
SF = iface.addVectorLayer('Path to your shape file.shp','SF_neighborhoods','ogr')
QgsProject.instance().addMapLayer(SF)
if SF.isValid():
    QgsProject.instance().addMapLayer(SF)
```

The code here is obtaining the polygon shapes for the neighborhoods in the layer we're calling SF_neighborhoods. If the vector is valid, this code will add the layer to the canvas. The `.isValid():` is a program check to verify that the input entered is correct.

You have now loaded a layer onto the canvas from the Python Console. The color of the vector layer is random, but you can modify the attributes:

```
renderer = SF.renderer()
symbol = renderer.symbol()
symbol.setColor(QColor('pink'))
```

Next, you'll want to include the names of the neighborhoods. Select Layer Properties and format the Labels to update the map with the names. In Figure 3-6, you can see how to adjust the Font, Style (bold), Size, Color, and Opacity.

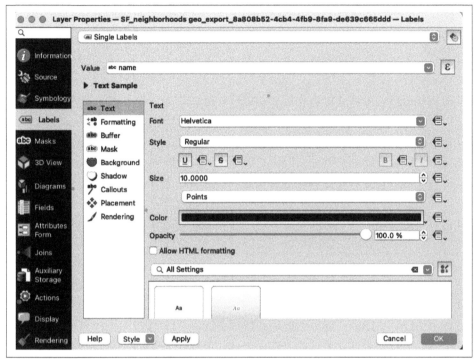

Figure 3-6. Layer properties: adding labels to a map

The labels are added to the canvas (Figure 3-7) when you select Apply and OK.

Load in the remaining layers. The quickest way is to use the addVectorLayer() method as follows:

```
vlayer = iface.addVectorLayer("Path to SF Urban Tree Canopy.shp", "SF_TREES",
"ogr")
if not vlayer:
  print("Layer failed to load!")

vlayer = iface.addVectorLayer("Path to Equity Strategy Neighborhoods.shp",
"ESN layer", "ogr")
if not vlayer:
  print("Layer failed to load!")
```

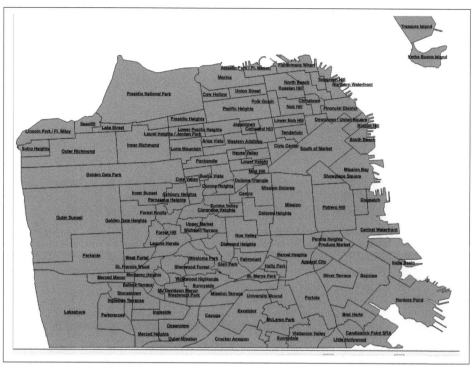

Figure 3-7. SF neighborhood layer from QGIS, with labels

Earlier, you used the method `QgsProject.instance().addMapLayer(SF)` to set the active layer. Here, you are adding the vector layer, `iface.addVectorLayer`.

Working with Layer Panels

Before we view the final map, I want to rewind a little to show you how we got here. On your canvas you have three vector layers.

Click on View in the menu bar and scroll down to Panels (Figure 3-8) to see a nested menu of Panel options. You can dock these options onto your canvas to make them accessible while you are building a visualization (map) and interacting with the features and underlying data.

Figure 3-8. Types of panels available in QGIS

To dock panels, you can drag them on top of one another to save space and move between them as needed by selecting the panel. If you decide you prefer to undock them, click the stacked image next to the X, and one by one they will undock. I suggest docking Layers >> Layer Styling >> Browser >> Processing Toolbox. As you begin working, you can remove or add panels simply by returning to your canvas or by clicking View in the menu and scrolling to Panels, then checking or unchecking options.

The panels are a type of widget. You can use them to provide inputs and visibility, digitize coordinates, perform statistical analysis, and add data sources, to name a few examples. See the QGIS User Guide (*https://oreil.ly/QWQF1*) for details on all the different options. I will walk you through the panels I use in almost all of the map projects I create.

In Figure 3-9, you can see that the layers in the upper left window have been moved into a hierarchy that allows each to be visible. For example, the labels for the neighborhoods are at the highest level, so they aren't buried beneath the other features or the polygons.

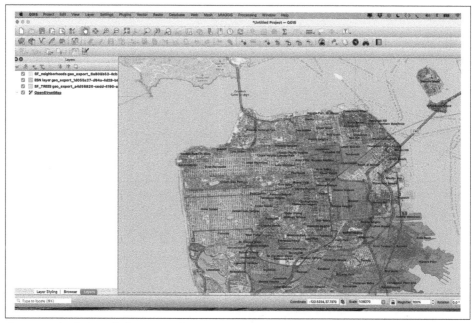

Figure 3-9. Editing the layers on the map canvas

In the Layers panel, you can also adjust opacity and colors. You want to be certain that the Equity Strategy Neighborhoods data is visible but not masking the SF Urban Tree Canopy beneath. Select the panels and arrange them on the canvas.

Now that you have been introduced to writing Python code in the console, the QGIS Cheat sheet for PyQGIS (*https://oreil.ly/ yXc82*) has a list of code snippets for you. Feel free to experiment with them and decide which might introduce efficiencies in your workflow.

Addressing the Research Question

Now let's address our original research question. How *does* the tree-cover data line up with low-income neighborhoods? Can you get a sense of the answer by looking at the map?

At a glance, the tree cover seems sparse in many of the low-income areas we've identified. Performing additional calculations of tree-cover density would be the next step in quantifying what we observe in the maps. Selecting one of the neighborhoods reveals the underlying data (Figure 3-10). We won't come to any formal conclusions here, but I encourage you to see how deep you can go.

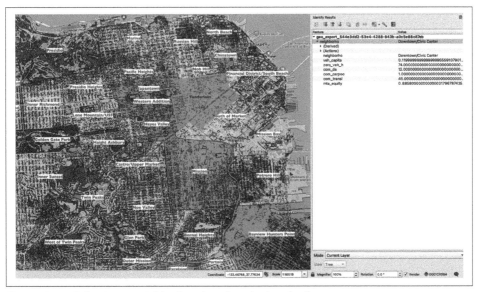

Figure 3-10. Selecting a neighborhood and exploring the features

This exercise helped you get familiar with the QGIS workspace and the Python plug-in. You learned how to work with layers and add labels. You also learned about finding existing datasets that can serve as a proxy for something you want to measure, and you got a sense of how aligning two datasets on a map can help address research questions.

Web Feature Service: Identifying Environmental Threats in Massachusetts

In the next example, you will explore another open dataset. The Massachusetts state government's MassGIS Bureau of Geographic Information (*https://oreil.ly/SNfaJ*) provides a GIS tool called MassMapper (*https://oreil.ly/JickV*). One of the many data layers provided within this tool is called Areas of Critical Environmental Concern (ACECs) (*https://oreil.ly/9Z3zd*). This layer is developed and maintained by the Massachusetts Department of Conservation and Recreation (DCR). According to the site, this data provides information about places in Massachusetts selected for the "quality, uniqueness, and significance of their natural and cultural resources." Using the ACECs data, we can build a map and see how these areas might be affected by local features, such as the density of nearby roadways and proximity to wetlands or airports.

Accessing the Data

There are many ways to connect to data in QGIS, so here we'll explore a new one.

The Web Feature Service (WFS) specification is a type of provider key that allows access to geographic features, with geometry and attributes available for your queries and analyses. Connect to the data by selecting Layer >> Data Source Manager or by clicking the icon in the toolbar. Select WFS/OGC API - Features. You can also scroll vertically in your Browser panel. Click on the WFS/OGC API, and the pop-up in Figure 3-11 will become visible.

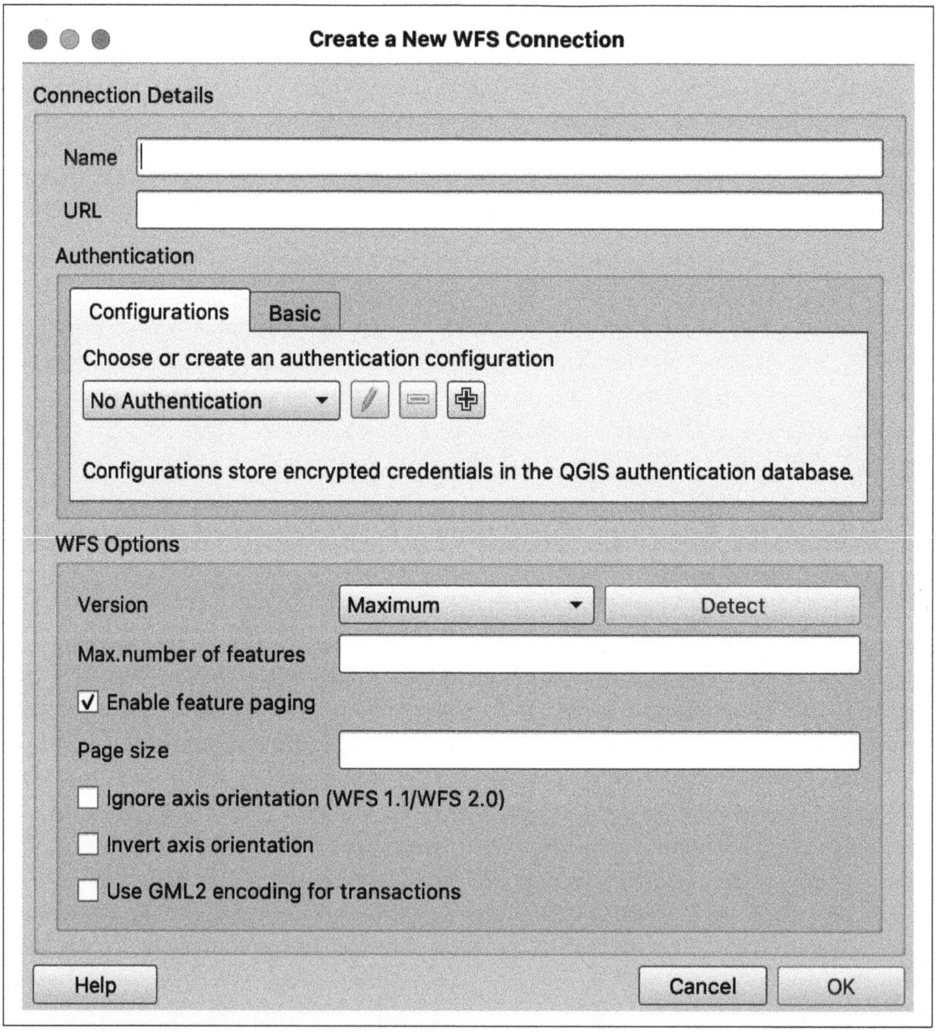

Figure 3-11. Creating a new WFS connection

Connect to the service by entering the URL *http://giswebservices.mass-gis.state.ma.us/geoserver/wfs* into the dialog field (Figure 3-12). You'll be prompted to create a name for the connection: use MASS_Sample and click OK. The defaults are fine.

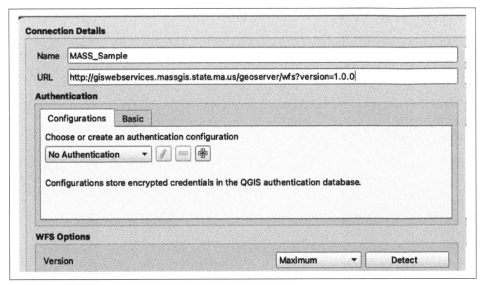

Figure 3-12. WFS connection details

The new connection appears in Server Connections. Click Connect and you will see the list of available layers, as pictured in Figure 3-13.

Scroll down to find and select the following layers, then click Add:

- GISDATA.AIRPORTS_PT (locations of airports)
- GISDATA.BM2_CH_BIOMAP2_WETLANDS (locations of wetlands)
- GISDATA.ACECS_POLY (areas of critical environmental concern)
- GISDATA.CENSUS2010TIGERROADS_ARC (locations of roads)

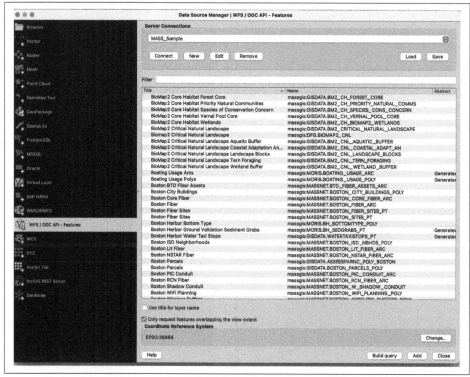

Figure 3-13. Data Source Manager WFS connection

Discovering Attributes

Enter the following code into the console and hit run (the green arrow):

```
active_layer = iface.activeLayer()
iface.showAttributeTable(active_layer)
iface.showLayerProperties(active_layer)
```

This creates a reference to the active layer, in this case, GISDATA.AIRPORTS_PT. Once you create this reference, you can access the properties of the layer.

Now you can explore a few of the options or access them directly in the console. First, select Open Attribute Table to view the attributes associated with the layer. You can do the same for any layer you want to explore.

As you type code into the console, you may notice that suggestions appear, as shown in Figure 3-14.

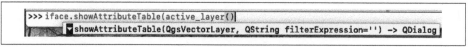

Figure 3-14. Automatic suggestions in the QGIS Python Console

---

Let's look at the snippet that appears when opening the attribute table:

```
showAttributeTable(QgsVectorLayer, QString filterExpression='') -> QDialog
```

The yellow box contains the arguments you can supply to the `showAttributeTable` method and what will be delivered. Here you are calling the `QgsVectorLayer` object and a string that represents a filter expression, which you can use to filter the data by identifying a specific field, as shown in Figure 3-15. The `filterExpression` is not required, which is indicated by the empty string: `=''`. If left empty, it simply provides the default, an empty string in this case. The output `QDialog` constructs the actual attribute table. This is just a brief sample; to learn more about the full functionalities of this particular class, see the `QgisInterface` documentation (*https://oreil.ly/70feB*).

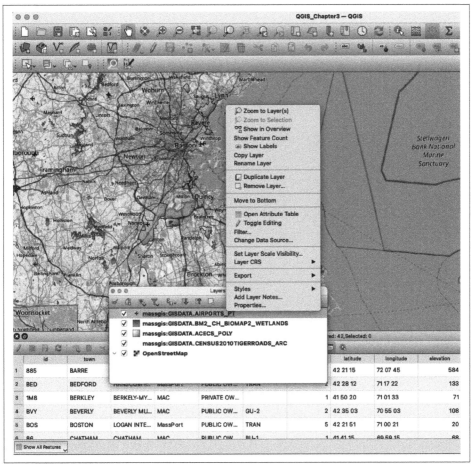

Figure 3-15. Layers and attribute table

Working with Iterators

Next, let's find the airports. Writing the following code directly into the console identifies the layer and feature count of the towns where airports are located:

```
layer = iface.activeLayer()
print(type(layer))

<class 'qgis._core.QgsVectorLayer'>

print(layer.sourceName())
massgis:GISDATA.AIRPORTS_PT
layer.featureCount()
42
```

And here is a snippet of the output:

```
BEDFORD
BEVERLY
BOSTON
HANSON
HOPEDALE
...
```

An iterator is a type of Python object that contains items that can be iterated upon. Iterators give you a window into a larger dataset, one element at a time. They are similar to other objects, like lists, but with a key difference: when you create an iterator, you don't store all the items in memory. The iterator loads a single item at a time and then fetches the next item when asked for it. This makes iterators very efficient. They can read large amounts of data without having to read the entire dataset. QGIS implements iterators for many different object types.

Lists in Python

Lists are mutable. You can add, delete, or change elements within a list. In Python, lists are defined with comma-separated values between square brackets [ ]. You can access each item by its position, or index. In programming, the counting starts from 0, not 1. So the first item has an index of 0, the second item an index of 1, and so on.

In the following code, the result of calling `layer.getFeatures()` is an iterator; the `next()` function is a manual iterator; and lists, tuples, and strings are iterables:

```
layer = iface.activeLayer()
features = layer.getFeatures()
f = next(features)
print(f.attributes())
f = next(features)
print(f.attributes())
```

This outputs:

```
['BED', 'BEDFORD', 'HANSCOM FIELD', 'MassPort', 'PUBLIC OWNED AP', 'TRAN',
2, '42 28 12', '71 17 22', 133, 7001.0, 5106, 150, 150, 'ASPHALT-GROOVED',
'ASPHALT', 'HIRLS', 'MIRLS', 'FAA', 'N', ' ', 'RELIEVER W/ COMMERCIAL SERVICE',
'Y']
['BVY', 'BEVERLY', 'BEVERLY MUNICIPAL AIRPORT', 'MAC', 'PUBLIC OWNED AP', 'GU-2',
2, '42 35 03', '70 55 03', 108, 5001.0, 4637, 150, 100, 'ASPHALT', 'ASPHALT',
'MIRLS', 'MIRLS', 'CONTRACT', 'Y', ' ', 'RELIEVER GA', 'N']
```

If you want a list of all the layers in your map, use the following code:

```
for layer in QgsProject.instance().mapLayers().values():
    print(layer.name())
```

It then outputs:

```
OpenStreetMap
massgis:GISDATA.ACECS_POLY
massgis:GISDATA.AIRPORTS_PT
massgis:GISDATA.BM2_CH_BIOMAP2_WETLANDS
massgis:GISDATA.CENSUS2010TIGERROADS_ARC
```

Selecting Layer Properties will also open up tables, as shown in Figure 3-16. Running the code `layer.featureCount()` earlier showed 42 features total, in this case, airports (the active layer) plus the 23 fields (columns) listed in the Layer Properties panel.

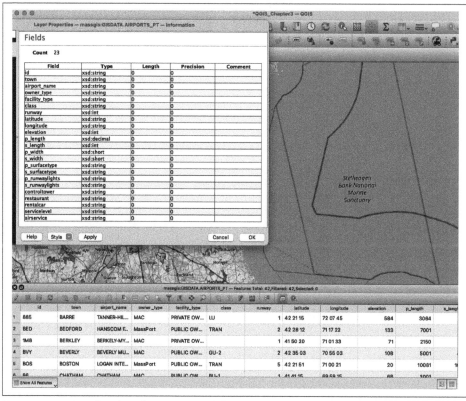

Figure 3-16. Attributes and properties of the selected layer, GISDATA.AIRPORTS_PT

Adding sample code to the Python Console will generate the information for the map layer. When running code a single line at a time, you can use the console and type your code at the >>> prompt. When running multiline scripts, you will want to use the editor: the panel on the right in Figure 3-17. Line numbers populate as the code runs. You will see this in the final example in the chapter.

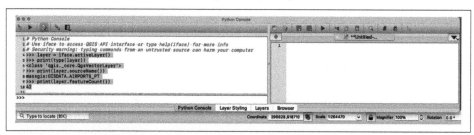

Figure 3-17. Running Python code in the console to add data to the canvas

Layer Styling

The layers can now be modified to make features visible and clear so that your audience will understand them. The colors generated are often random, so I like to customize them using Layer Styling.

Now you can view the map (Figure 3-18) and explore airport locations, roadways, and wetlands and how they relate to ACECs areas. Where do you see potential for these features to affect the health of these environmentally sensitive areas? What other features might you want to look for in answering this research question?

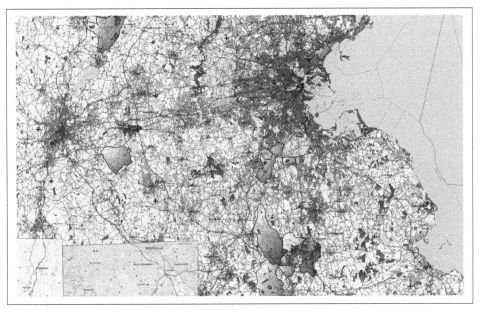

Figure 3-18. Map of Massachusetts ACECs areas

This exercise gave you some more practice with QGIS layers and introduced you to layer styling. You also learned how to work with iterators and discover attributes.

Using Processing Algorithms in the Python Console

So far, you've been learning about the capabilities of both the Python Console and QGIS in geospatial analysis. You are gaining familiarity with an information system that combines geographic data with robust software for managing, analyzing, and visualizing that data. By now, you've likely discovered even more plug-ins and powerful tools in the Processing Toolbox. You've got a great start on using Python for engaging with a system like QGIS.

From here on, you can continue building your expertise by making use of help functions, QGIS documentation, and other resources. You'll practice that in the following exercise. For this exercise, our research question will be: *which cities are located along the Amazonas River?* Perhaps you are hoping to focus your attention on how flooding or agricultural runoff affect local communities. The exercise is an opportunity to try filtering and analyzing a large open source dataset. Explore the attribute tables and customize a query!

One of the biggest advantages of Python scripting is that your tasks can be flexible and easy to reproduce. In the final section of this chapter, you will learn how to use *processing algorithms*: algorithms that let you save your scripts and chain them together. Adapting existing scripts is one of the best ways to learn how Python operates. You'll be working with a script adapted from Anita Graser[1] to build a workflow in the Python console. This workflow will use three processing algorithms:

- native:extractbyexpression
- native:buffer
- native:extractbylocation

As we did in Chapter 2, we'll use data from Natural Earth Data. The canvas in Figure 3-19 shows two kinds of data for Brazil: populated places and rivers and lakes. Download GeoPackage (*https://oreil.ly/IMccr*), if you have not done so already. Select GeoPackage from Data Source Manager to upload the files.

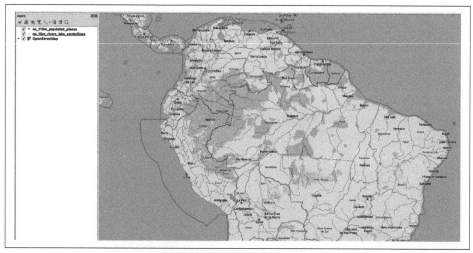

Figure 3-19. Exploring the area of interest in the map canvas

---

1 This workflow is adapted from Anita Graser's undated blog post "PyQGIS 101: Chaining Processing Tools" (*https://oreil.ly/RdQxx*), a tutorial for working with processing tools.

Working with Algorithms

Before we start, you need to understand where these algorithms originate and how to work with the different parameters. Write the following code in your Python Console to get started:

```python
from qgis import processing
```

When working with functions, it's important to call algorithms by name so that they execute reliably. The `QgsProcessingRegistry` (*https://oreil.ly/yXOkX*) will list the algorithms correctly. Registries are where you can access algorithms, parameters, and different outputs. You can access the registry by writing this code:

```python
for alg in QgsApplication.processingRegistry().algorithms():
    print(alg.id(), "->", alg.displayName())
```

Algorithms with names starting with `native` are processing algorithms that still port to C++ and impart a speed advantage over other algorithms.

You can scroll through the long list in the output (excerpted below) to see the others, but you can also access them directly with a little bit of code:

```
native:atlaslayouttomultiplepdf -> Export atlas layout as PDF (multiple files)
native:atlaslayouttopdf -> Export atlas layout as PDF (single file)
native:batchnominatimgeocoder -> Batch Nominatim geocoder
native:bookmarkstolayer -> Convert spatial bookmarks to layer
native:boundary -> Boundary
native:boundingboxes -> Bounding boxes
native:buffer -> Buffer
...
```

I counted more than two thousand algorithms when I ran this code. Each algorithm needs certain parameters to execute its code successfully. Use the parameter `processing.algorithmHelp("algorithmID")` with the name of the algorithm to output the characteristics of the syntax. This will help familiarize you with writing the code.

First, let's look at the script we'll use. Don't run it yet! Before you do, I'll break down each piece:

```python
buffered_amazonas = processing.run("native:buffer",
    {'INPUT':amazonas,'DISTANCE':buffer_distance,'SEGMENTS':5,'END_CAP_STYLE':0,
    'JOIN_STYLE':0,'MITER_LIMIT':2,'DISSOLVE':False,'OUTPUT':'memory:'}
    )['OUTPUT']my_gpkg = "/Users/bonnymcclain/Downloads/natural_earth_vector/
        packages/natural_earth_vector.gpkg"
rivers = '{}|layername=ne_110m_rivers_lake_centerlines'.format(my_gpkg)
places ='{}|layername=ne_110m_populated_places'.format(my_gpkg)
expression = "name = 'Amazonas'"
amazonas = processing.run("native:extractbyexpression",
    {'INPUT':rivers,'EXPRESSION':expression,'OUTPUT':'memory:'}
    )['OUTPUT']
```

```
buffer_distance = 0.1 #degrees


places_along_amazonas = processing.run("native:extractbylocation",
    {'INPUT':places,'PREDICATE':[0],'INTERSECT':buffered_amazonas,'OUTPUT':
        'memory:'}
    )['OUTPUT']

QgsProject.instance().addMapLayer(places_along_amazonas)

for feature in places_along_amazonas.getFeatures():
    print(feature["name"])
```

Notice that it runs your three algorithms in order. First, you'll extract by expression. The result is a layer that is stored in the `amazonas` variable and passed as input into the next algorithm. The buffer algorithm creates buffers that intersect with buffered river centerlines. The distance from the river, for example, should be within a certain distance of the city. The buffer sets this distance. Next, `extractbylocation` pulls your list of cities along the river.

Extract by Expression

Here's how the `algorithmHelp()` function describes native:extractbyexpression:

> This algorithm creates a new vector layer that only contains matching features from an input layer. The criteria for adding features to the resulting layer is based on a QGIS expression.

Take a moment to run `processing.algorithmHelp("native:extractbyexpression")` and read the rest now. It will provide important details and describe the output you can anticipate. If you don't list a parameter, the default will be applied and output is sent to memory.

Here is the segment of the script that uses this algorithm, so you can examine it in more detail:

```
amazonas = processing.run("native:extractbyexpression",
    {'INPUT':rivers,'EXPRESSION':expression,'OUTPUT':'memory:'}
    )['OUTPUT']
buffer_distance = 0.1 #degrees
```

Buffer

Next, run:

```
processing.algorithmHelp("native:buffer")
```

Let's see what the help documentation (*https://oreil.ly/pNmzd*) tells us about native:buffer:

This algorithm computes a buffer area for all the features in an input layer, using a fixed or dynamic distance.

The segments parameter controls the number of line segments to use to approximate a quarter circle when creating rounded offsets.

The end cap style parameter controls how line endings are handled in the buffer.

The join style parameter specifies whether round, miter or beveled joins should be used when offsetting corners in a line.

The miter limit parameter is only applicable for miter join styles, and controls the maximum distance from the offset curve to use when creating a mitered join.

Read the full output for the information you need to understand the algorithm and the inputs and outputs needed. In particular, you need to understand the parameters so you can include them in your query. You will see information for INPUT, DISTANCE, SEGMENTS, END_CAP_STYLE, JOIN_STYLE, MITER_LIMIT, DISSOLVE, and OUTPUT.

Here is the script segment that includes this algorithm:

```
buffered_amazonas = processing.run("native:buffer",
    {'INPUT':amazonas,'DISTANCE':buffer_distance,'SEGMENTS':5,'END_CAP_STYLE':0,
    'JOIN_STYLE':0,'MITER_LIMIT':2,'DISSOLVE':False,'OUTPUT':
    'memory:'}
    )['OUTPUT']
```

What is this code doing? What can you tell from the syntax here? From reading the output of the algorithmHelp function, you can see that the location vector is now the input for the native.buffer. The buffer will be interested only in data within the buffer range. For example, the rivers and cities should be within 10 kilometers.

Extract by Location

Our third algorithm is native:extractbylocation. Next, we'll *intersect* the buffered Amazonas data–that is, compare the two sets of data we now have–to combine the rivers data with the places data. This will include the cities along the river we are interested in.

First, run:

```
processing.algorithmHelp("native:extractbylocation")
```

The help documentation again contains important information about the parameters.

Here is the section of our script that uses extract by location:

```
places_along_amazonas = processing.run("native:extractbylocation",
    {'INPUT':places,'PREDICATE':[0],'INTERSECT':buffered_amazonas,'OUTPUT':
    'memory:'}
    )['OUTPUT']
```

The QGIS GUI contains similar tools in the Processing Toolbox functions (Figure 3-20), but you need to run the processes separately and locate the suitable input and output parameters in output (which will resemble that in Figure 3-21).

```
processing.algorithmHelp("native:extractbylocation")
```

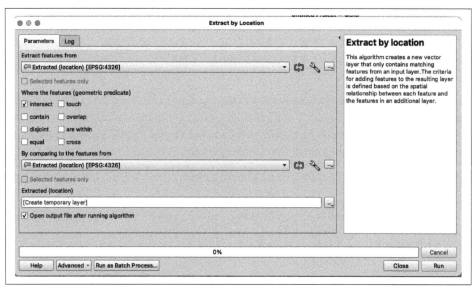

Figure 3-20. Algorithms in the QGIS GUI

Figure 3-21. The input and output parameters buried in the QGIS GUI output

Now that you've examined it, go ahead and run the script in the console. The output, shown in Figure 3-22, is a map of cities along the Amazonas river in Brazil. This exercise is an example of how creating algorithms to run on large datasets reinforces efficient and repeatable workflows. Now you can query the data with your own ideas. Use the attribute tables available for each dataset to note how the variables or column headings are listed. Saving your Python scripts will allow you to create templates to practice and update as your curiosity grows!

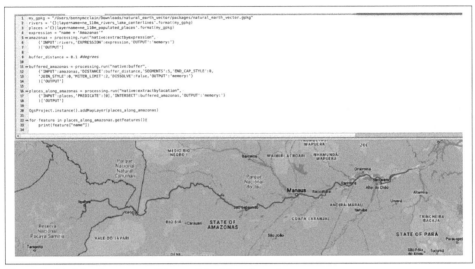

Figure 3-22. The Amazonas River with cities along its route

Summary

This chapter showed you how to download QGIS, customize your workspace, and upload data from different provider keys, both within the Python Console and directly from the Browser. You also learned to customize your maps to increase their legibility and started becoming familiar with the canvas and layer styling. You were introduced to the Processing Tools. These are complex, but splitting the code into small sections and learning about the syntax is often perfect practice for gaining more advanced skills. You learned how to chain three algorithms together and store results in a variable, to be used as input.

In coming chapters, you will continue to build your Python skills by interacting with additional tools like Google Earth Engine and ArcGIS.

## CHAPTER 4

Geospatial Analytics in the Cloud: Google Earth Engine and Other Tools

How do you access geospatial data? Although data professionals with enterprise accounts may not think about the limitations of personal computing and relying on open source data, the rest of us often work within limits. Geospatial analysis in the cloud has narrowed the divide, since that means we no longer need to store large volumes of data locally. Never before has the general public had open source access to geospatial data on such a global scale. This chapter will show you where to find data for exploration and learning.

Space programs in the US and around the world have gathered data from satellites and sensors for decades, but only recently have we had the capacity to manipulate that data in real time for analysis. The USGS hosts EarthExplorer (*https://oreil.ly/ OnxdN*) (Landsat), and the Copernicus Open Access Hub (*https://oreil.ly/gnY7c*) provides data from European Space Agency (ESA) Sentinel satellites. Landsat high-resolution satellite images enable us to evaluate and measure environmental change, understand the impact of climate science and agricultural practices, and respond to natural disasters across time and space, to name a few examples. The advent of free satellite images has enabled decision makers from economically challenged areas across the world to bring insights into view and focus on solutions.

Spatial analysis includes methods and tools applied to location data, in which the results vary based on the location or frame analyzing the object. It is essentially "location-specific" analysis. This can be as simple as locating the nearest subway station or asking how many green spaces or parks are in a community, or as complex as revealing patterns in transportation accessibility or health outcomes. *Spatial algorithms* are a method of solving a problem by listing and executing sequential instructions integrated with geographic properties, used for analysis, modeling, and prediction.

GIS solve spatial problems that rely on location information like latitude, longitude, and projection. Spatial information answers "where" questions: where on the Earth's surface did something occur?

Imagine, for example, stepping out of your hotel on 41st and Madison Avenue in Manhattan. You search in your mapping app for where you might purchase a coat, since the weather is dramatically colder than you anticipated. Instantly, the locations of apparel stores populate your screen.

Or on the marketing side, say you work for an outdoor provision company, producing top-of-the-line outerwear for the discerning customer. You could use geospatial information to answer questions like: Where do your potential customers live, visit, or travel? Would a potential retail location nearby be a profitable marketing decision? How far would potential customers travel? What is the mean income within each of the locations you are considering? These *where* components exist in retail and commercial environments, the military, climate science, and health care, to name a few examples.

Attributes are another important component of spatially referenced data. *Spatial attributes* are bounded in space; these could include a community boundary or infrastructure, such as a roadway or metro station, usually represented by a polygon. Spatially referenced data can also have nonspatial attributes, such as the income of residents in a certain location, and can provide context for the location intelligence.

The *I* in GIS is increasingly being stored in the cloud. Today your laptop can access petabytes of information made available by geospatial analytics processing services in the cloud. This chapter will explore one of those services: Google Earth Engine (GEE) (*https://oreil.ly/ukyb0*).

In 2007, Jim Gray, a computer scientist at Microsoft until he was lost at sea later that year, was quite prescient when he said: "For data analysis, one possibility is to move the data to you, but the other possibility is to move your query to the data. You can either move your questions or the data. Often it turns out to be more efficient to move the questions than to move the data." That's the basic principle behind doing geospatial analytics in the cloud.

In this chapter, you'll use GEE to perform a variety of tasks associated with geographic properties in spatial environments. We'll also take a quick look at another tool that integrates with Python: Leafmap. By the end of the chapter, you'll have enough familiarity with these interfaces to follow along with later chapters and eventually launch your own independent project.

Google Earth Engine Setup

But first, you'll need to create your work environments. The Jupyter Notebooks for this chapter are available on GitHub (*https://oreil.ly/SbS0R*). You can open them up and follow along or experiment with the code and explore separately when time permits. The instructions for installing the necessary packages and resources will be covered as well.

The GEE archive contains more than 60 *petabytes* of satellite imagery and remote sensing and geospatial data—all freely available, preprocessed, and easy to access. Imagine trying to download all that to your laptop! GEE's algorithms allow the public to create interactive applications or data products in the cloud. You just need to apply for a free Google Earth Engine account (*https://oreil.ly/xVOCN*) (which comes with 250 gigabytes of storage) and authenticate within either the terminal or notebook when you are granted access. Follow these steps:

```
To authorize access needed by Earth Engine, open the following URL
in a web browser and follow the instructions:
https://accounts.google.com/o/oauth2/auth?client_id=xxx
The authorization workflow will generate a code, which you should paste
in the box below
Enter verification code: code will be here

Successfully saved authorization token.
```

GEE will send you a unique link and verification code. Paste the code into the box and hit Enter.

Using the GEE Console and geemap

The GEE console is a quick resource for locating images and running the code. But Python isn't GEE's native language: the GEE code editor is designed for writing and executing scripts in JavaScript. Its Javascript API has a robust IDE, extensive documentation, and interactive visualization functionality, and none of that is natively available for Python. To access the full spectrum of interactivity in a Python environment, you will need to use geemap (*https://geemap.org*), a Python package for interacting with GEE created by Dr. Qiusheng Wu (*https://oreil.ly/bGQq1*).

Fortunately, you can use the extensive GEE catalog to locate and visualize data with a single click, with limited or no JavaScript expertise. You can find your way around the interface and generate maps simply by scrolling through the Scripts tab. Each code script allows you to run JavaScript code and generate maps. But if you're seeking autonomy to build your own maps and engage interactively, you'll want to use geemap. The GEE catalog (pictured in Figure 4-1) contains useful information you will need when deciding how to interact with data in geemap.

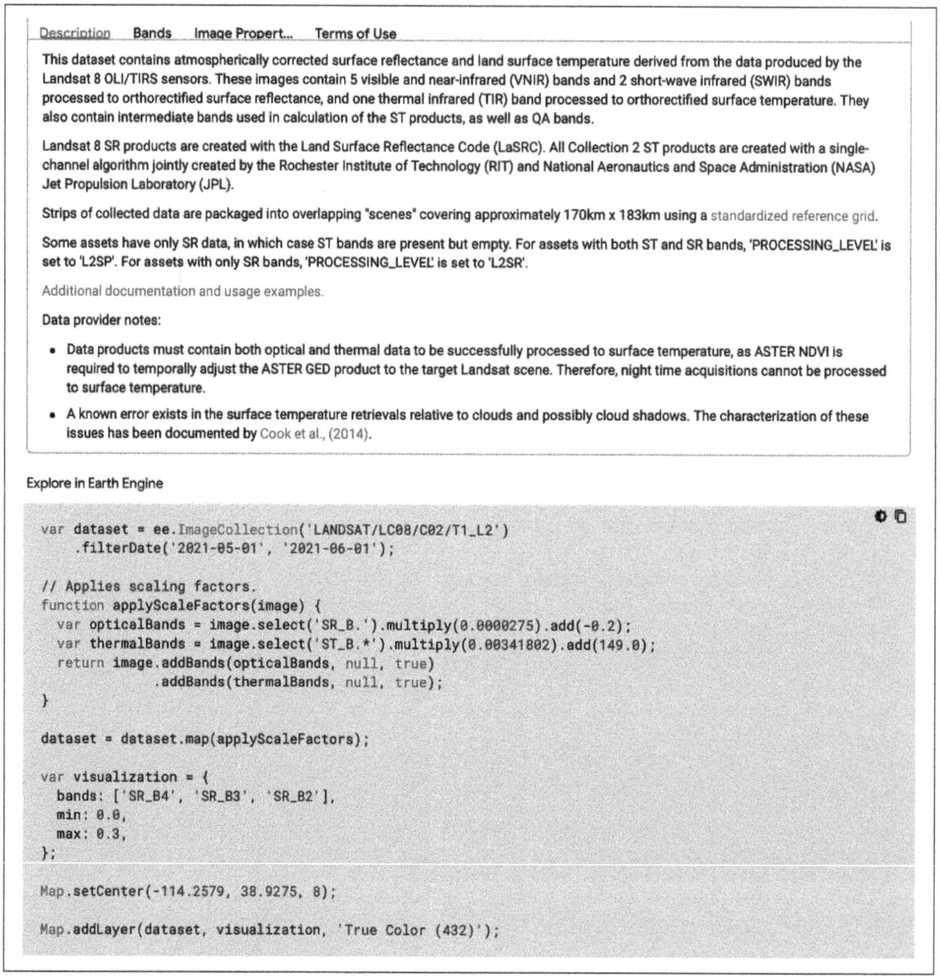

This dataset contains atmospherically corrected surface reflectance and land surface temperature derived from the data produced by the Landsat 8 OLI/TIRS sensors. These images contain 5 visible and near-infrared (VNIR) bands and 2 short-wave infrared (SWIR) bands processed to orthorectified surface reflectance, and one thermal infrared (TIR) band processed to orthorectified surface temperature. They also contain intermediate bands used in calculation of the ST products, as well as QA bands.

Landsat 8 SR products are created with the Land Surface Reflectance Code (LaSRC). All Collection 2 ST products are created with a single-channel algorithm jointly created by the Rochester Institute of Technology (RIT) and National Aeronautics and Space Administration (NASA) Jet Propulsion Laboratory (JPL).

Strips of collected data are packaged into overlapping "scenes" covering approximately 170km x 183km using a standardized reference grid.

Some assets have only SR data, in which case ST bands are present but empty. For assets with both ST and SR bands, 'PROCESSING_LEVEL' is set to 'L2SP'. For assets with only SR bands, 'PROCESSING_LEVEL' is set to 'L2SR'.

Additional documentation and usage examples.

Data provider notes:

- Data products must contain both optical and thermal data to be successfully processed to surface temperature, as ASTER NDVI is required to temporally adjust the ASTER GED product to the target Landsat scene. Therefore, night time acquisitions cannot be processed to surface temperature.

- A known error exists in the surface temperature retrievals relative to clouds and possibly cloud shadows. The characterization of these issues has been documented by Cook et al., (2014).

Explore in Earth Engine

```
var dataset = ee.ImageCollection('LANDSAT/LC08/C02/T1_L2')
    .filterDate('2021-05-01', '2021-06-01');

// Applies scaling factors.
function applyScaleFactors(image) {
  var opticalBands = image.select('SR_B.').multiply(0.0000275).add(-0.2);
  var thermalBands = image.select('ST_B.*').multiply(0.00341802).add(149.0);
  return image.addBands(opticalBands, null, true)
              .addBands(thermalBands, null, true);
}

dataset = dataset.map(applyScaleFactors);

var visualization = {
  bands: ['SR_B4', 'SR_B3', 'SR_B2'],
  min: 0.0,
  max: 0.3,
};

Map.setCenter(-114.2579, 38.9275, 8);

Map.addLayer(dataset, visualization, 'True Color (432)');
```

Figure 4-1. The GEE catalog

Look through the Earth Engine Data Catalog, find a dataset collection (*https://oreil.ly/yAow0*), and scroll down the page. At the bottom, you will notice that the JavaScript code is provided. Simply copy and paste it into the console, as shown in Figure 4-2.

Figure 4-2 shows what is generated when you paste the code into the console and select Run from the list of options in the center panel. For example, the data from Figure 4-1 generates USGS Landsat 8 Level 2, Collection 2, Tier 1, identified as `ee.ImageCollection("LANDSAT/LC08/C02/T1_L2")`.

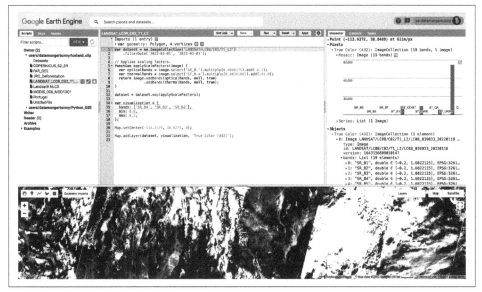

Figure 4-2. Google Earth Engine IDE

Let's learn how to generate GEE images using Python scripts in a Jupyter Notebook. geemap even has a tool that will convert JavaScript code to Python right in your Jupyter Notebook.

Jupyter Notebook is a separate entity from your Python environments. It was originally named for its ability to to interact with three different coding languages, Julia, Python, and R, but it has expanded well beyond its original capabilities. You have to tell the system which version of Python you want. The *kernel* is how the Notebook and Python communicate.

Installing geemap will create a console in a Notebook environment similar to what you see in the GEE console but with the Python API instead of JavaScript. Once you set up a Conda environment, you will be able to interact with GEE within a Jupyter Notebook. First, you will need to download the required packages and libraries.

Creating a Conda Environment

Anaconda (*https://www.anaconda.com*) is a popular platform-agnostic distribution manager for Python and other programming languages that installs and manages Conda packages. You could think of Anaconda as storage for all of your data science tools. Conda manages your packages and tools, allowing you to upload new tools as needed and to customize your work environment.

Conda packages are stored in the Anaconda repository or the cloud, so you don't need additional tools for installation. Conda allows you to make as many environments as you need with your preferred version of Python. You also have the option of downloading a leaner version of Anaconda called Miniconda (*https:// oreil.ly/nh0LE*), which I prefer, regardless of your operating system. Both are straightforward installations. I recommend the Miniconda installation instructions in this tutorial by Ted Petrou (*https://oreil.ly/nPTlh*).

Opening the Jupyter Notebook

Jupyter Notebooks are open source, interactive, web-based tools. They run in your browser and don't require any additional downloads. You can access the Jupyter Notebook for this chapter on GitHub (*https://oreil.ly/SbS0R*): the filename is *4 Geospatial Analytics in the Cloud*. You can find and configure the installed *nbextensions* in the file menu of your Notebook. These are handy plug-ins that add more functionality to your Jupyter Notebook environment.

Installing geemap and Other Packages

Once you've installed your Conda environment, you can open your terminal or command prompt to install geemap. Execute the following code line by line to activate your work environment. Here, I've named my geospatial environment **gee**:

```
conda create -n gee python=3.9
conda activate gee
conda install geemap -c conda-forge
conda install cartopy -c conda-forge
conda install jupyter_contrib_nbextensions -c conda-forge
jupyter contrib nbextension install --user
```

Notice that I specified the version of Python to include in the environment. I've done this because there are still some dependencies that aren't ready for the latest version of Python. That is one important reason why environments are useful: you will receive a warning if there are compatibility conflicts, and you can create an environment using the version that will avoid those conflicts.

This install also includes Cartopy, a Python package for geospatial data processing; `jupyter_contrib_nbextensions`, a package for expanded functionality; and `con trib_nbextensions`, which will add styles to the Jupyter configuration.

Now that you've installed the packages into your environment, whenever you open a new session, you will only need to run `import geemap` in a code cell. The environment is now visible when you activate, shown here as (**gee**):

```
(gee) MacBook-Pro-8:~ bonnymcclain$ conda list
```

This environment will contain all of the associated packages as well as their dependencies. You can create different environments (abbreviated as `env`) that include the dependencies and packages unique to each project.

The conda list command will show you which packages are installed in the active environment. This following is a snippet of what loads for me when I execute the command:

```
# packages in environment at /Users/bonnymcclain/opt/miniconda3/envs/geo:
#
# Name                        Version         Build           Channel
aiohttp                       3.7.4           py38h96a0964_0  conda-forge
anyio                         3.1.0           py38h50d1736_0  conda-forge
appnope                       0.1.2           py38h50d1736_1  conda-forge
argon2-cffi                   20.1.0          py38h5406a74_2  conda-forge
async-timeout                 3.0.1                  py_1000  conda-forge
async_generator               1.10                      py_0  conda-forge
attrs                         21.2.0          pyhd8ed1ab_0    conda-forge
backcall                      0.2.0           pyh9f0ad1d_0    conda-forge
backports                     1.0                       py_2  conda-forge
backports.functools_lru_cache 1.6.4           pyhd8ed1ab_0    conda-forge
beautifulsoup4                4.9.3           pyhb0f4dca_0    conda-forge
bleach                        3.3.0           pyh44b312d_0    conda-forge
bqplot                        0.12.27         pyhd8ed1ab_0    conda-forge
branca                        0.4.2           pyhd8ed1ab_0    conda-forge
brotlipy                      0.7.0           py38h5406a74_1001 conda-forge
bzip2                         1.0.8           h0d85af4_4      conda-forge
c-ares                        1.17.1          h0d85af4_1      conda-forge
ca-certificates               2020.12.5       h033912b_0      conda-forge
cachetools                    4.2.2           pyhd8ed1ab_0    conda-forge
cartopy                       0.19.0.post1    py38h4be4431_0  conda-forge
certifi                       2020.12.5       py38h50d1736_1  conda-forge
cffi                          1.14.5          py38ha97d567_0  conda-forge
chardet                       4.0.0           py38h50d1736_1  conda-forge
click                         8.0.1           py38h50d1736_0  conda-forge
colour                        0.1.5                     py_0  conda-forge
geemap                        0.8.16          pyhd8ed1ab_0    conda-forge
...
```

This is helpful in case your code throws an error due to a missing dependency. Run conda list; you should see the versions listed as well. Running conda env list will display any environments you already have installed.

I install a kernel (a part of the operating system running in your environment) for each environment that I activate:

```
conda install ipykernel
```

Now you can add the kernel to your environment—in this case, *<your environment name>* is gee:

```
python -m ipykernel install --user --name myenv --display-name
"<your environment name>"
```

Your local computer has to access files. The `import` statement will add the package as a Python *object* (that is, a collection of data and methods) into the currently running instance of the program.

Open your Terminal and write **jupyter notebook** into the console. A Notebook should open in your browser. You will need to import the required libraries into the Notebook. You can see them listed in the code shell. Recall that `os` allows you to access the operating system where you are running Python, `ee` is the Earth Engine library, and `geemap` allows you to interface via Python. You'll import these libraries using the `import` function:

```
import os
import ee
import geemap
#geemap.update_package()
```

The central component of a computer operating system is the *kernel*. The kernel is specific to each programming language, and the default kernel depends on what version of Python you are running in your Notebook.

You will need to restart the kernel for the update to take effect. Select Kernel from the menu and scroll to the option to rerun. You are now ready to begin working in the Notebook.

Note the hash symbol (#) in the last line of the previous code block. In Python code, the hash symbol denotes a *comment*, or a line of the code that won't run. When you want to run that line of code, delete the hash symbol. To make sure you are using an updated geemap package, *uncomment* that last line (that is, remove the # in the last row) before running the code. Once you update geemap, you can once again insert the hash, since you won't need to update the package every time you run the code. You may also add commented text to include any clarifying details. You will see this practice in many of the code blocks in this book.

Navigating geemap

I mentioned objects earlier. Python focuses on objects, instead of what you may be familiar with as *functions* in other programming languages, so it's known as an *object-oriented* programming language.

You may recall from earlier chapters that a *class* is like a blueprint of a building The building is the object, but many buildings can be built from a set of blueprints, right? The object in a specific line of code is an *instance* of the class much in the same way that a building is an *instance* of those blueprints. Creating the object is called *instantiation*.

The next code block is declaring an object instance, which I'm calling map, and defining the attributes and methods in geemap.Map(). You can set your variable to anything you would like, but be consistent. I suggest keeping it simple but informative and practical. You'll access the attributes of objects using the object name map. geemap defaults to a world map (as of this writing). If you prefer to center your map in a specific place, you can indicate where using latitude/longitude (lat/lon) coordinates and a zoom level. The following will center your map on the US:

```
map = geemap.Map(center=(40, -100), zoom=4)
map
```

Layers and Tools

Figure 4-3 shows the Layers and Tools menu on the far right of the map. Each layer of a map is actually its own database that holds collections of geographic data. These might include roads, buildings, streams, or lakes, all represented as collections of points, lines, and/or polygons in vector data or imagery represented from raster data. The Layers icon will show you the different layers in your map. You can change their opacity, toggle layers on and off, and examine other attributes. Enter the following code to render a map in your notebook:

```
map = geemap.Map(center =(40, -100), zoom=4)
map.basemap()
map
```

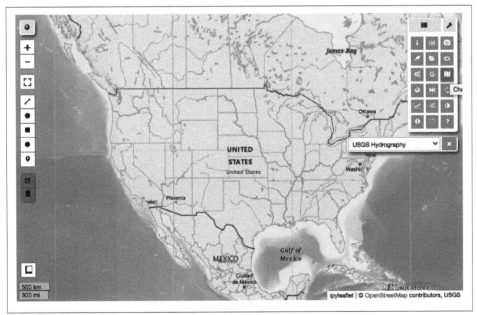

Figure 4-3. The Layers and Tools menu

Click the different options to see how they customize the map.

The globe icon (on the far left in Figure 4-4) is the search location/data function. This is where you can find data to load onto your map by entering a name and address or a set of lat/lon coordinates or by searching and importing data. We will explore more of these options as we build a few map layers, and I'll show you some shortcuts to help you navigate the mapping canvas.

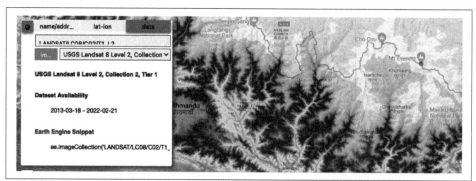

Figure 4-4. Searching location data using GEE asset ID (Landsat/LC08/C01) in the import window

You can access USGS Landsat maps by entering the search parameters in geemap. Select "data" and enter a few terms into the search bar. Again, exploring GEE may help to identify maps of interest. You can call them to your own project here by name. Once you select a Landsat collection and select the asset ID, click the import button (seen in blue in Figure 4-4).

To see available arguments and additional customization options, place your cursor within the parentheses of geemap.Map() and hit Shift + Tab. The text below the map in Figure 4-5 should now be visible. Here you can read information about available arguments and additional steps for further customization of the map.

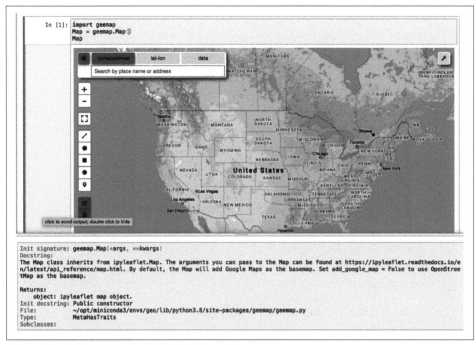

```
In [1]: import geemap
        Map = geemap.Map()
        Map
```

Figure 4-5. A basemap in GEE with docstring

Basemaps

Revisit Figure 4-3 for a moment, and look on the far right for a dropdown menu. This menu contains a dictionary of available *basemaps*: basic maps that serve as the foundation for your data exploration. The Jupyter Notebook lets you scroll through available basemaps without writing code.

Depending on your data question or the nature of the data, you may want to show different geospatial information. For example, if you are interested in showing *hydrography* (the physical characteristics and navigability of bodies of water), you likely won't select a basemap that depicts major roadways and highways. Basemaps are stored as raster or vector tiles for expediency and performance. The basemap dictionary interacts with a *TileLayer*, allowing connections with map services such as NASA's Global Imagery Browse Services (GIBS) (*https://oreil.ly/AOwFm*) and Open-StreetMap (*https://oreil.ly/1Ti4J*).

The geemap package brings all the analytic functionality of GEE into *ipyleaflet,* an interactive library that brings mapping into your notebook, allowing the dynamic updates you see in the maps as you update locations and zoom levels. The default map in geemap is the Google Maps global view. You're going to use OSM as your basemap next, so run:

```
add_google_map = False
```

Advanced users have the option of creating their own TileLayer, but a variety of other default basemaps are freely available in the ipyleaflet map and basemaps (*https:// oreil.ly/HJJbT*) resource.

Now that you know how to load a map into your Notebook, let's get brave and start experimenting. The goal is for you to get curious and feel comfortable navigating the Jupyter Notebook and selecting different tools.

Exploring the Landsat 9 Image Collection

We have been working with Landsat data, so let's look at the Landsat 9 data, which was first released in early 2022 and still rolling out as of this writing. To see how much of the dataset is available, run the following code:

```
collection = ee.ImageCollection('LANDSAT/LC09/C02/T1_L2')
print(collection.size().getInfo())
```

This outputs: 106919.

The collection includes 106,919 images—and is still increasing!

For comparison, the Landsat/LC08/C02/T1_L2 collection contains 1,351,632 images. By the time this book is published, the number of Landsat 9 images will be vastly larger. You can calculate the median value of all matching bands to reduce the size of the image collection:

```
median = collection.median()
```

Working with Spectral Bands

As you learned in Chapter 1, *spectral bands* are like bins of different types of light. Reflected light is captured as bands of light energy in a range of different wavelengths or colors. Think of the electromagnetic spectrum. Each section of the spectrum is actually a band. The information about bands in this section is intended to highlight where to locate the data to enter into the code to access the correct information. The bands (*https://oreil.ly/NN6Eq*) collected by Landsat 8 apply to Landsat 9. You will need this data to apply *scaling factors*, or comparisons of linear distances that adjust for distortion of areas and angles based on the map projection (also covered in Chapter 1). Remember, the Earth is shaped as an ellipsoid, not a perfect sphere! We derive scaling factors from the Scale and the Offset, as shown in Figure 4-6.

Description	Bands	Image Propert...	Terms of Use				

Resolution
30 meters

Bands

Name	Units	Min	Max	Scale	Offset	Wavelength	Description
SR_B1		1	65455	2.75e-05	-0.2	0.435-0.451 μm	Band 1 (ultra blue, coastal aerosol) surface reflectance
SR_B2		1	65455	2.75e-05	-0.2	0.452-0.512 μm	Band 2 (blue) surface reflectance
SR_B3		1	65455	2.75e-05	-0.2	0.533-0.590 μm	Band 3 (green) surface reflectance
SR_B4		1	65455	2.75e-05	-0.2	0.636-0.673 μm	Band 4 (red) surface reflectance
SR_B5		1	65455	2.75e-05	-0.2	0.851-0.879 μm	Band 5 (near infrared) surface reflectance
SR_B6		1	65455	2.75e-05	-0.2	1.566-1.651 μm	Band 6 (shortwave infrared 1) surface reflectance
SR_B7		1	65455	2.75e-05	-0.2	2.107-2.294 μm	Band 7 (shortwave infrared 2) surface reflectance
SR_QA_AEROSOL							Aerosol attributes
⊕ Bitmask for SR_QA_AEROSOL							
ST_B10	Kelvin	0	65535	0.00341802	149	10.60-11.19 μm	Band 10 surface temperature. If 'PROCESSING_LEVEL' is set to 'L2SR', this band is fully masked out.
ST_ATRAN		0	10000	0.0001			Atmospheric Transmittance. If 'PROCESSING_LEVEL' is set to 'L2SR', this band is fully masked out.
ST_CDIST	km	0	24000	0.01			Pixel distance to cloud. If 'PROCESSING_LEVEL' is set to 'L2SR', this band is fully masked out.
ST_DRAD	W/(m^2*sr*um)/DN	0	28000	0.001			Downwelled Radiance. If 'PROCESSING_LEVEL' is set to 'L2SR', this band is fully masked out.
ST_EMIS		0	10000	0.0001			Emissivity of Band 10 estimated from ASTER GED. If 'PROCESSING_LEVEL' is set to 'L2SR', this band is fully masked out.

Figure 4-6. Band characteristics of Landsat 8/9 data

The USGS provides guidance on which spectral bands are best for different types of research. You can learn more about the science (*https://oreil.ly/vTrFF*) and explore common Landsat band combinations (*https://oreil.ly/HdXC0*).

You'll import *ee.ImageCollection* into your Jupyter Notebook and add it as a data layer to your map. You'll then create a composite image from all of the images. This will yield the median value of the spectral bands.

In Python, we define functions by the keyword def. In the following code, the function name is apply_scale_factors, followed by the parameter (image):

```python
def apply_scale_factors(image):
    opticalBands = image.select('SR_B.*').multiply(0.0000275).add(-0.2)
    thermalBands = image.select('ST_B.*').multiply(0.00341802).add(149.0)
    return image.addBands(opticalBands, None, True).addBands(thermalBands, None,
    True)
```

The asterisk (*) tells the function that you want to select multiple bands that meet the defined search requirements. Landsat's sensors are the Operational Land Imager (OLI) and Thermal Infrared Sensor (TIRS). The OLI produces spectral bands 1 through 9, and TIRS consists of two thermal bands: SR_B and ST_B.

The colon (:) signals where the function body begins. Inside the function body, which is indented, the `return` statement determines the value to be returned. After the function definition is complete, calling the function with an argument returns a value:

```
dataset = apply_scale_factors(median)
```

To understand why you would want to pick and choose certain bands, think of them as having a spectral signature. Natural color bands use SR_B4 for red, SR_B3 for green, and SR_B2 for blue. Green indicates healthy vegetation, brown is less healthy, whitish gray typically indicates urban features, and water will appear dark blue or black.

The near-infrared (NIR) composite uses NIR (SR_B5), red (SR_B4), and green (SR_B3). Areas in red have better vegetation health, dark areas are water, and urban areas are white. So include these as your visualization parameters:

```
vis_natural = {
    'bands': ['SR_B4', 'SR_B3', 'SR_B2'],
    'min': 0.0,
    'max': 0.3,
}

vis_nir = {
    'bands': ['SR_B5', 'SR_B4', 'SR_B3'],
    'min': 0.0,
    'max': 0.3,
}
```

Now add them as data layers to your map:

```
Map.addLayer(dataset, vis_natural, 'True color (432)')
Map.addLayer(dataset, vis_nir, 'Color infrared (543)')
Map
```

In Figure 4-7, I toggled the infrared layer off so you can see the other bands more clearly. There appears to be cloud cover as well. Landsat 9 resamples every 16 days, so it will look different when you view it.

Figure 4-7. Different band combinations of Landsat 8/9

If you hover your cursor over the toolbar icon, you will see the Layers menu appear. You can change the opacity of any maps and deselect any layers you don't want to view in the Layers menu. You can also click the gear icon to explore attributes. You can also specify the minimum and maximum values to display. Stretching the data spreads the pixel values, and you can experiment with different values. Your data will show the range of the bands, and you can decide which values you want to display:

```
vis_params = [
    {'bands': ['SR_B4', 'SR_B3', 'SR_B2'], 'min': 0, 'max': 0.3},
    {'bands': ['SR_B5', 'SR_B4', 'SR_B3'], 'min': 0, 'max': 0.3},
    {'bands': ['SR_B7', 'SR_B6', 'SR_B4'], 'min': 0, 'max': 0.3},
    {'bands': ['SR_B6', 'SR_B5', 'SR_B2'], 'min': 0, 'max': 0.3},
]
```

To add labels for these layers, create a list of labels:

```
labels = [
    'Natural Color (4, 3, 2)',
    'Color Infrared (5, 4, 3)',
    'Short-Wave Infrared (7, 6 4)',
    'Agriculture (6, 5, 2)',
]
```

Then assign a label to each layer:

```
geemap.linked_maps(
    rows=2,
    cols=2,
    height="400px",
    center=[-3.4653, -62.2159],
    zoom=4,
    ee_objects=[dataset],
    vis_params=vis_params,
    labels=labels,
    label_position="topright",
)
```

Examining two more parameters in Figure 4-8, you can also see shortwave infrared. Here, darker green indicates dense vegetation, urban areas are shown in blue, healthy vegetation is green, and bare earth is magenta.

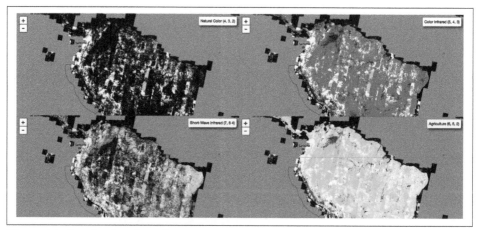

Figure 4-8. Landsat band combinations

Let's apply your introduction to GEE and geemap to begin exploring.

The National Land Cover Database Basemap

The National Land Cover Database (NLCD) (*https://oreil.ly/xNJul*) tracks land cover in the US. It is freely available in the Earth Engine Data Catalog (*https://oreil.ly/1IuYF*) and is updated every five years. *Land cover* data includes spatial reference and land surface characteristics, such as tree canopy cover (which we explored in the last chapter), impervious surfaces, and additional patterns of biodiversity and climate changes. *Impervious land cover* means nonnatural surfaces, such as asphalt, concrete, or other manmade layers, that limit the natural penetration of rainwater into soil. This information can help predict which areas may be more prone to flooding during heavy rains.

In this section, you'll use NLCD data to perform a Landsat-based examination of the imperviousness data layer (for urban classes) and of decision-tree classification (for the rest). We won't be doing a full activity here, just a quick orientation, but I encourage you to explore more.

Accessing the Data

Navigate to the Earth Engine Data Catalog (*https://oreil.ly/oWZcr*) and scroll to NLCD_Releases/2019_REL/NLCD or the National Land Cover Database, as shown in Figure 4-9. Earlier we noted that you can simply add this data to the map, but there are a few more options here that I want to show you. Copy the JavaScript code and place it on your clipboard.

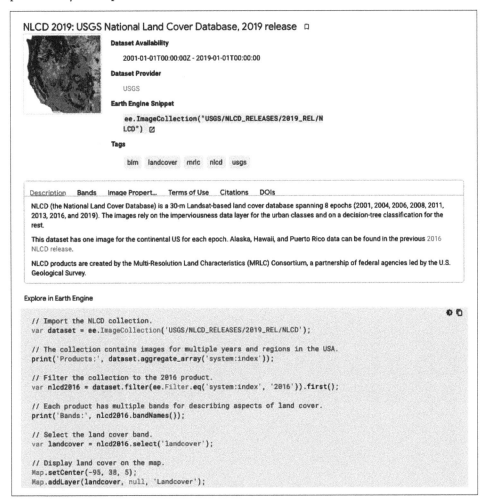

Figure 4-9. The Earth Engine Data Catalog

The NLCD catalog provides a wealth of information, including date ranges for collection, the data source, an image snippet, a data description, information about the multispectral bands, and image properties.

In geemap, generate a default map of the world:

```
map = geemap.Map()
map
```

Next, select the convert JavaScript icon. The box shown in Figure 4-9 will pop up. Paste the JavaScript code from the catalog into the box. Follow the instructions in the code comments that populate in the pop-up shown in Figure 4-10:

```javascript
// Import the NLCD collection.
var dataset = ee.ImageCollection('USGS/NLCD_RELEASES/2019_REL/NLCD');

// The collection contains images for multiple years and regions in the USA.
print('Products:', dataset.aggregate_array('system:index'));

// Filter the collection to the 2016 product.
var nlcd2016 = dataset.filter(ee.Filter.eq('system:index', '2016')).first();

// Each product has multiple bands for describing aspects of land cover.
print('Bands:', nlcd2016.bandNames());

// Select the land cover band.
var landcover = nlcd2016.select('landcover');

// Display land cover on the map.
Map.setCenter(-95, 38, 5);
Map.addLayer(landcover, null, 'Landcover');
```

Once you hit Convert, you will see the code update from JavaScript to Python, as shown in Figure 4-10.

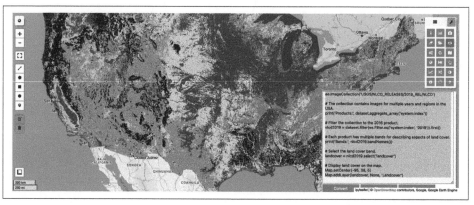

Figure 4-10. Using geemap to convert a script from JavaScript to Python

If the code does not update into a new cell in your Jupyter Notebook, you can cut and paste it into a new cell and run the cell. The image will now appear as your map.

Now let's include the default NLCD legend. Select the landcover layer. To discover which legends are available as defaults, run the `builtin_legend` function:

```python
legends = geemap.builtin_legends
for legend in legends:
    print(legend)
```

The NLCD's legend will be listed as an option. Select it to add it to your map.

Building a Custom Legend

While the NLCD offers a built-in legend option, many datasets do not—and even when they do, these legends don't always offer exactly what you need. Thus, it's helpful to be able to create your own map legend. Let's look at how to do that now.

The classes in a dataset usually correspond to the categories you'd want in a legend. Fortunately, you can convert a class table to a legend.

If your data is from the GEE data catalog, you can find a class table there. Then use the following code (or find this code cell in your Jupyter Notebook) and copy the text from the class table into it:

```
map = geemap.Map()

legend_dict = {
    '11 Open Water': '466b9f',
    '12 Perennial Ice/Snow': 'd1def8',
    '21 Developed, Open Space': 'dec5c5',
    '22 Developed, Low Intensity': 'd99282',
    '23 Developed, Medium Intensity': 'eb0000',
    '24 Developed High Intensity': 'ab0000',
    '31 Barren Land (Rock/Sand/Clay)': 'b3ac9f',
    '41 Deciduous Forest': '68ab5f',
    '42 Evergreen Forest': '1c5f2c',
    '43 Mixed Forest': 'b5c58f',
    '51 Dwarf Scrub': 'af963c',
    '52 Shrub/Scrub': 'ccb879',
    '71 Grassland/Herbaceous': 'dfdfc2',
    '72 Sedge/Herbaceous': 'd1d182',
    '73 Lichens': 'a3cc51',
    '74 Moss': '82ba9e',
    '81 Pasture/Hay': 'dcd939',
    '82 Cultivated Crops': 'ab6c28',
    '90 Woody Wetlands': 'b8d9eb',
    '95 Emergent Herbaceous Wetlands': '6c9fb8'
}

landcover = ee.Image('USGS/NLCD/NLCD2019').select('landcover')
Map.addLayer(landcover, {}, 'NLCD Land Cover')

Map.add_legend(legend_title="NLCD Land Cover Classification",
    legend_dict=legend_dict)
Map
```

You can find more info on building and customizing legends manually in the geemap documentation (*https://geemap.org*).

Now you can explore your map and dig deeper into your areas of interest. What questions do you want to ask of this map? Take some time to explore. There are many different ways to customize your maps with a broad selection of tools!

The GEE catalog (*https://oreil.ly/3Q25Z*) is extensive. As you explore different databases and datasets using the skills you've learned here, you will be able to work with raster and vector data as well as upload your own data sources. A list of handy additional functions in geemap is available on the geemap GitHub page (*https://gee map.org/usage*). However, I'd also like to introduce you to an alternative to GEE.

Leafmap: An Alternative to Google Earth Engine

Visualizing geospatial data outside of GEE does not have to be limiting! If you don't have access to a GEE account or aren't interested in working with GEE, consider using Leafmap. Leafmap is a Python package that lets you visualize interactive geospatial data in your Jupyter Notebook environment. It is based on the geemap package you have already experienced, but as you will see in this section, Leafmap provides access to geospatial data outside the GEE platform. Its GUI reduces the amount of coding you need to do. It has a variety of open source packages at its core.

Leafmap works with many different plotting backends, including ipyleaflet. (A *backend*, in this context, is internal code that runs on a server and receives client requests.) Users don't see backends, but they are always operating nonetheless.

You can access the Jupyter Notebook Leafmap with the GitHub link (*https://oreil.ly/9ADWy*). Follow the Leafmap documentation (*https://oreil.ly/f2Ztx*) for specific installation instructions depending on your version of Python. (If you aren't sure what version you have, enter **python** in the terminal, and it will output the number of the version you have installed. This is important to remember in case you run into issues with your installation of packages.)

You can set up a new environment to work with Leafmap. I originally created the Conda environment shown in the following code with Python 3.8, but it is likely to work with later versions. I named this environment **geo** because it is running in a different version of Python:

```
conda create -n geo python=3.8
conda activate geo
conda install geopandas
conda install leafmap -c conda-forge
conda install mamba -c conda-forge
mamba install leafmap xarray_leaflet -c conda-forge
conda install jupyter_contrib_nbextensions -c conda-forge
pip install keplergl
```

As before, to open the Notebook, type **jupyter notebook** and hit Enter. Now enter the following code into the Notebook to reveal something similar to Figure 4-11:

```
from ipyleaflet import *
m = Map(center=[48.8566, 2.3522], zoom=10, height=600, widescreen=False,
basemaps=basemaps.Stamen.Terrain)
m
```

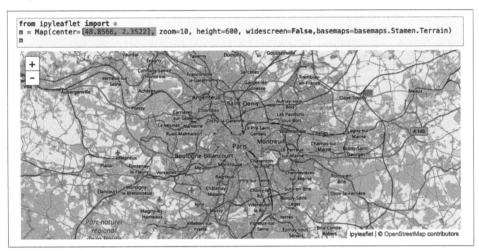

Figure 4-11. Installing basemaps in Leafmap

Changing the basemap is as easy as placing your cursor inside the basemap parentheses and selecting Tab on the keyboard. Figure 4-12 shows the options that become available. Esri is the selected basemap here, but you can scroll up and down until you find a suitable one. Be sure to explore. Once you type **Esri**, options will populate.

Figure 4-12. Changing basemaps in Leafmap

Another useful tool is the ability to preset your zoom levels. When you run the cell in your Notebook, you will have the option of sliding between different zoom levels:

```
m.interact(zoom=(5,10,1))
```

Figure 4-13 shows the output.

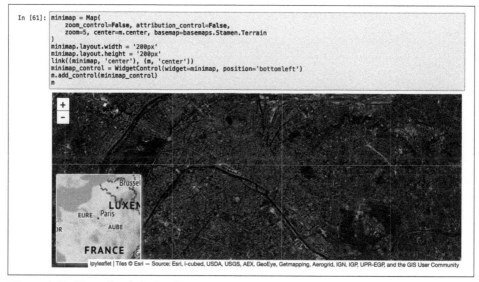

```
In [61]: minimap = Map(
             zoom_control=False, attribution_control=False,
             zoom=5, center=m.center, basemap=basemaps.Stamen.Terrain
         )
         minimap.layout.width = '200px'
         minimap.layout.height = '200px'
         link((minimap, 'center'), (m, 'center'))
         minimap_control = WidgetControl(widget=minimap, position='bottomleft')
         m.add_control(minimap_control)
         m
```

Figure 4-13. Zoom levels in Leafmap

You can also provide a reference by inserting a minimap into your larger map, as shown in Figure 4-14. To do so, enter the following code:

```
minimap = Map(
    zoom_control=False, attribution_control=False,
    zoom=5, center=m.center, basemap=basemaps.Stamen.Terrain
)
minimap.layout.width = '200px'
minimap.layout.height = '200px'
link((minimap, 'center'), (m, 'center'))
minimap_control = WidgetControl(widget=minimap, position='bottomleft')
m.add_control(minimap_control)
```

The minimap shown in Figure 4-14 will appear, helping users stay oriented in a larger context.

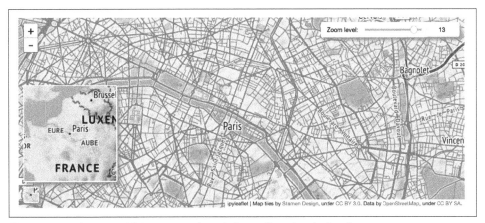

Figure 4-14. A map within a map: the minimap function in Leafmap

Summary

This chapter explored Google Earth Engine and some related tools, libraries, and packages that you can use to answer geospatial questions, and it introduced you to an alternative tool, Leafmap. This chapter and its associated Notebooks will be a handy reference for the projects you'll do in the next chapter. You have rendered visualizations and created maps on the canvas. Next, you will begin analyzing these relationships and exploring tools to do some advanced analysis of your geospatial data.

## CHAPTER 5

OpenStreetMap: Accessing Geospatial Data with OSMnx

OpenStreetMap (OSM) (*https://oreil.ly/UWkLH*) is an editable geographic database of the entire world built by volunteers with an auspicious goal: to create geographic data and distribute it to all of us for free. You have interacted with OSM already on your smartphone if you use GPS for directions—or any location-enabled device. Python offers a package called OSMnx (*https://oreil.ly/CR2vV*) that lets urban planners and a wide variety of other users create street networks and interact with and analyze otherwise "hidden" geographic information. You can find walkable, drivable, or bikeable urban networks for your own personal use or for research, such as to study characteristics of urban environments. Robust analytics reveal infrastructure frameworks that disclose inefficiencies when analyzing the network and interrelated nature of roadways, for example.

Personally, I think street networks are works of art. But their real use, which is perhaps underutilized, is adding geometric shapes to your built infrastructure. You can add buildings (hospitals, schools, and grocery stores, for example), parks, and other dataframes categorized as edges, buildings, and areas. The term *building* is defined loosely, and there is a Wiki for building tags (*https://oreil.ly/ZuB1n*). A great resource to find place names is OSM Nominatim (*https://oreil.ly/tmo4M*). You can also add points of interest, elevations, and much more.

At first glance, OSMnx may seem a bit technical and complicated. But as you build street networks with OSM in this chapter, you'll learn to apply and interact with this powerful, customizable package.

A Conceptual Model of OpenStreetMap

The OSM database (*https://oreil.ly/pPT7s*) contains more than 80,000 tag keys and 600 relation types, so you can define the appropriate granularity for your map. In this chapter, you will learn how to access these features, customize them, add them to your map, and perform analyses. For example, you can identify a location and explore the distribution of agricultural, commercial, or residential land use. (A snapshot of different land-use values you can access by querying the database is shown in Figure 5-2, later in this section.) Understanding the distribution of land use in an area may be important if you are interested in how much flooding occurs there or how to calculate storm runoff. The most popular tags in the database as of this writing include data about buildings, highways, land surface, and waterways.

OSM's structure has a few rules, but they are relatively straightforward. For example, nodes can be arbitrary (*hashable*) Python objects, with both nodes and edges as optional key/value attributes. To view these optional values in their entirety, head over to either OSM map features (*https://oreil.ly/XDg2a*) or tag info (*https://oreil.ly/oOkoG*) (or follow along in the text).

Is Hashable the Same as Immutable?

You may recall immutable objects as being values that do not change. Their value is set, and you cannot edit them—think tuples, like lists of days of the week, months of the year, and so forth.

If something is hashable, you can call `hash()`, and its value will not change during its lifetime. This is important when creating `key=value`.

You can put a mutable object (like a list) inside a tuple. The tuple is still immutable, but you can change the list inside it, so it's not hashable.

Tags

OSM applies a *tag* made up of a key-value pair. You will see these defined in the code cells we will write in this chapter. Their format is a *key* and its corresponding *value*, formatted as `key=value`. For example, if your key for a location is set to `highway` and the value is `residential`, you can tell that this is a road where people live.

Here are a few examples of tags:

```
surface=asphalt
highway=residential
building=commercial
```

Depending on what you're looking for, this level of granularity may be useful in your data exploration. I have used these tags to explore impervious surfaces in urban settings, for example. Impervious surfaces tend to trap heat, and places with a lot of them tend to have higher rates of flooding—important information when comparing the characteristics of different neighborhoods. As another example, you might simply want a map of all the buildings in a neighborhood. The default value buildings="all" will then be uploaded. We'll look at the utility of including tags in the code examples and sample Notebook for this chapter.

OSMnx extracts OSM data and creates a routable Python NetworkX (*https://net workx.github.io*) object for working with complex networks. This is needed to convert edges and nodes from OSM, as shown in Figure 5-1, to links and junctions character-istic of routable networks. These routable maps are a network with features like travel time, speed limit, and shortest distance between locations allowing routing.

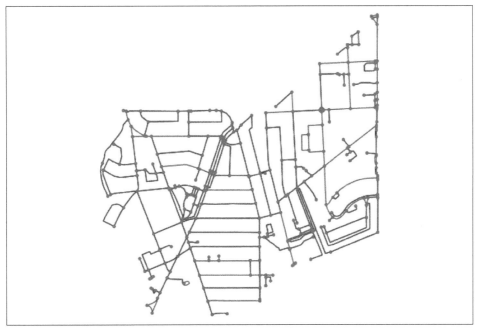

Figure 5-1. A map of the walkable street network in Manhattan's Chelsea neighborhood

OSMnx automates several mapping capabilities, including downloads of geometries like states, cities, neighborhoods, building footprints, customized street networks, and topology. In addition, OSMnx has built-in capabilities for analysis based on the foundation of graph theory. Simply defined, *graph theory* represents connections between elements and their locations within spatial networks, including both nodes and edges. The NetworkX integration allows you to customize nodes and edges to hold a variety of data, images, graph algorithms, and network structures for analysis.

Instead of solely geographic coordinates, vast amounts of information are stored for each location. We will begin exploring the types of data shortly. First, here's a brief introduction to what's under the hood.

Multidigraphs

Figure 5-1 shows edges between nodes. This is a preview of what we will build later in this chapter. The nodes are corners connected by street segments, which are weighted by length in this example. (You could also weight them by travel time, for example). This is one way to calculate the shortest paths between two places.

The data you are going to request in the next section will be in the format of a NetworkX multidigraph. *Multidigraphs* are abstract representations of objects or elements with multiple edges between the same nodes. They are directional, reflecting, for example, whether traffic on a specific city street is one-way or flows in both directions. Digraphs have edges that point from node to node, but not necessarily in both directions. Once you have multiple edges in parallel, you have a multidigraph.

OSM is a *wiki*, an open source, editable geographic database. Its conceptual model includes *nodes* that define points in space, *ways* defining linear features and boundaries, and *relations* that explain how nodes and ways work together. Think of a *node* as like a coordinate defined by latitude and longitude, with the *way* representing a list of nodes (called a *polyline*), or the boundaries of a polygon.

You might initially think of these spatial networks as simple planes, but there are plenty of nonplanar structures such as bridges, tunnels, and a variety of grade-separated structures like expressways, on- and off-ramps, and overpasses. You can evaluate these using topological measures and *metric structures*, or length and area described in spatial units (such as meters). The wiki description defines the relationship between the land-use tag and the values available to be assigned to a key, as shown in Figure 5-2.

Geographical data questions are often defined based on a specific location and usually have real-world implications. They often involve things like land use, road surfaces, number and types of buildings, or locations of community facilities, such as museums, bars, or internet access. For example, your data question might be something like:

- How many supermarkets are there in the seventh ward of Washington, DC?
- How many green spaces or parks are located in a particular neighborhood?
- How walkable is Chicago, Illinois?

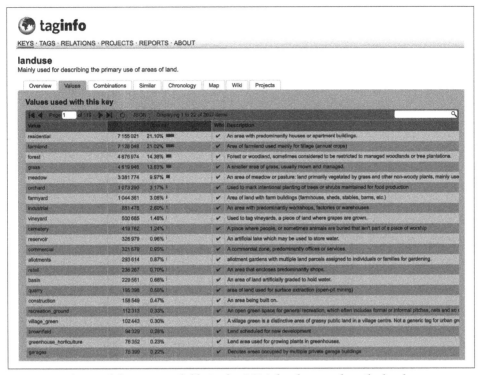

Figure 5-2. Some of the tags available in the OSM database to describe land use

First, let's install OSM.

Installing OSMnx

Install OSMnx to a unique environment within your terminal. As I've noted before, I create different environments as a simple way to address dependencies and updates. Another lesson learned the hard way: if you need to update the package, delete and reinstall it into your environment. I often delete environments after specific projects because they can take up significant space on your hard drive if you create them willy-nilly and let them sit there once no longer needed. This is easier if you name them with specificity. This seems to be the only way to be certain you have all of the dependencies also updated. If you receive a message in your code that a function has been deprecated, return to the user reference files (*https://oreil.ly/MZLFQ*) for clarification.

In your terminal, enter:

```
conda create -name OSM
conda activate OSM
conda install -c conda-forge osmnx
```

```
conda install -c conda-forge matplotlib
conda install -c conda-forge descartes
conda install -c conda-forge shapely
conda install -c conda-forge jupyter
conda install -c conda-forge jupyter_contrib_nbextensions
```

You'll be using OSMnx to retrieve the OSM data and matplotlib library to interact with and visualize your data.

If you want to see the packages you've installed alongside OSMnx, after `conda activate` env, enter **conda list** into the terminal. If you forget the name of your environment, you can write **conda env list** and view all of them. Here, I've created a unique environment called `OSM`:

```
(ox) MacBook-Pro-8:~ bonnymcclain$ conda list
# packages in environment at /Users/bonnymcclain/opt/miniconda3/envs/OSM:
```

When you are ready to launch a Notebook, type **jupyter notebook** into your terminal, and a Notebook will open.

Choosing a Location

To pass `place_name`, OSMnx uses OSM Nominatim API. The Nominatim Documentation (*https://oreil.ly/AZvAG*) includes an API reference guide to search for geocoding (see Figure 5-3). If you enter incorrect information (for example, if you're seeking a neighborhood by name but misspell it), you will likely get an error. Open Nominatim (*https://oreil.ly/AP9Sd*), and you will be taken to a debugging interface for the search engine that will look for the place you are requesting. I start there to make sure I am querying correctly.

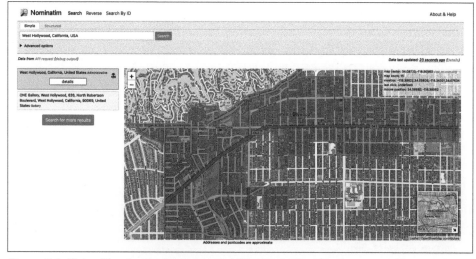

Figure 5-3. Using Nominatim to query an API request for `place_name`

Although you can add a place name as a string, it is best to check Nominatim and see how the location is listed to avoid conflicting or incorrect data. For example, I was looking for Brentwood, California, and got an error, so I ran it through Nominatim. It turned out that the boundaries were different for Brentwood as a neighborhood and for Brentwood's administrative boundaries—and then there's Brentwood Heights. I needed to make a minor adjustment in how I referred to the neighborhood.

This will make more sense when we begin working with these relationships. Refer to the OSMnx documentation (*https://oreil.ly/ZY8Nr*) for more, but you can also query directly inside of your code cell.

Let's say you want a detailed street map of Los Angeles. In the following code, you'll begin by importing the required packages: osmnx and matplotlib. Include place_name and set it equal to the desired location (within quotation marks in code). If you're looking at something more specific than city and state, the more detail you include, the better. Run the following code (it may take some time):

```
import osmnx as ox
import networkx as nx
import matplotlib.pyplot as plt #installs with osmnx

# download/model a street network for some city then visualize it
G = ox.graph_from_place("Los Angeles,California,United States",
network_type="drive")
fig, ax = ox.plot_graph(G,figsize=(20,20),bgcolor='#FFFFFF',
    node_color='black', node_size=0)
```

Functions are written in the format ox.module_name.function_name(). The previous code snippet references the osmnx.graph module and the graph_from_place function. Most functions can be called with ox.function_name() only. This will retrieve the geocoded information from Los Angeles,California,United States, along with drivable-street-network data within the boundaries (Figure 5-4).

Go ahead and pick a smaller location if you prefer. Larger cities are resource intensive and may take a few minutes to load. Later in the chapter we will explore Culver City, California.

Place your cursor inside the parenthesis in your code where the ☞ icon is pointing. Here is the snippet isolated for clarity:

```
G = ox.graph_from_place(☞"Los Angeles,California")
```

While your cursor is inside the parentheses, select Tab + Shift on your keyboard.

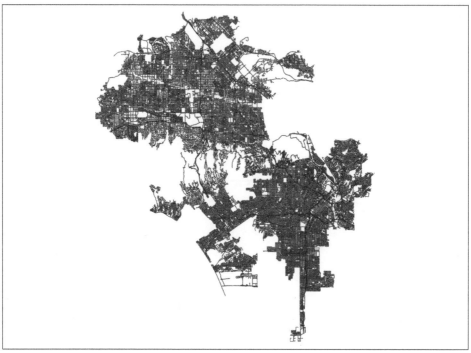

Figure 5-4. Street networks in the municipality of Los Angeles, generated by OSMnx

Let's see what parameters are possible!

For example, Figure 5-5 displays the characteristics of the signature. This is a list of the arguments, and they need to match the arguments listed in the signature of the function. Try the different options and see how the map changes.

```
1  G = ox.graph_from_place(place_name, clean_periphery=False,network_type="all_private")

Signature:
ox.graph_from_place(
    query,
    network_type='all_private',
    simplify=True,
    retain_all=False,
    truncate_by_edge=False,
    which_result=None,
    buffer_dist=None,
    clean_periphery=True,
```

Figure 5-5. Discovering arguments available within functions

The `network_type` options for the function shown in Figure 5-5 are:

`drive`
Drivable public streets (not service roads)

`drive_service`
Drivable public streets, including service roads

`walk`
All streets and paths that pedestrians can use (ignores one-way directionality by always connecting adjacent nodes with reciprocal directed edges)

`bike`
All streets and paths that cyclists can use

`all`
All (nonprivate) OSM streets and paths

`all_private`
All OSM streets and paths, including private access

Understanding Arguments and Parameters

Have you ever heard the expression, "Nobody is coming to save you"? I am convinced it was first uttered by a data scientist. When your code won't execute (and, trust me, that isn't an "if" so much as a "when"), you will need to go read the user documentation. That might be documentation about the package itself or perhaps a GitHub repository. Features and arguments are deprecated and updated with newer versions. Knowing how to find them is an important skill to develop.

If you scroll down past the parameters, you will notice customizations of these variables in the code snippets. I can't include every function and associated argument here, but let's take some examples. Select the inverted chevron in the upper right corner to display the contents of the box (^, as shown in Figure 5-5). Scroll down and you will see a section called "Docstring."

Docstrings provide quick information about a class, function, or method. I look here for hints on how to formulate my query. When you need additional information about a function, return your cursor inside the parentheses and select Tab + Shift. Here is an example of the type of documentation available:

```
Docstring:
Create graph from OSM within the boundaries of some geocodable place(s).

The query must be geocodable and OSM must have polygon boundaries for the geocode
result. If OSM does not have a polygon for this place, you can instead get its
street network using the graph_from_address function, which geocodes the place
name to a point and gets the network within some distance of that point.
```

If OSM does have polygon boundaries for this place but you're not finding it,
try to vary the query string, pass in a structured query dict, or vary the
which_result argument to use a different geocode result. If you know the OSM ID
of the place, you can retrieve its boundary polygon using the geocode_to_gdf
function, then pass it to the graph_from_polygon function.

As you continue to scroll down in the interface, you will see a "Parameters" section.
This section provides information about data types and how to complete the query.
Not all parameters are needed, depending on your query. If you leave some parameters out, the defaults will be displayed.

Take another look at the code snippet you used to generate the Los Angeles map in
Figure 5-4. This map requires entering information for the network_type parameter:
in this case, network_type="drive". The function takes a string (hence the quotation
marks), and the options are listed and must be within parentheses. Here they are, as
listed in the documentation:

```
Parameters
----------
query : string or dict or list
    the query or queries to geocode to get place boundary polygon(s)
network_type : string {"all_private", "all", "bike", "drive", "drive_service",
"walk"}
    what type of street network to get if custom_filter is None
simplify : bool
    if True, simplify graph topology with the `simplify_graph` function
retain_all : bool
    if True, return the entire graph even if it is not connected.
    otherwise, retain only the largest weakly connected component.
truncate_by_edge : bool
    if True, retain nodes outside boundary polygon if at least one of
    node's neighbors is within the polygon
which_result : int
    which geocoding result to use. if None, auto-select the first
    (Multi)Polygon or raise an error if OSM doesn't return one.
buffer_dist : float
    distance to buffer around the place geometry, in meters
clean_periphery : bool
    if True, buffer 500m to get a graph larger than requested, then
    simplify, then truncate it to requested spatial boundaries
custom_filter : string
    a custom ways filter to be used instead of the network_type
    presets  e.g., '["power"~"line"]' or
    '["highway"~"motorway|trunk"]'. Also pass in a network_type that
    is in settings.bidirectional_network_types if you want graph
    to be fully bi-directional.
```

Calculating Travel Times

A popular type of data question involves travel time: *how long will it take to get from point A to point B?*

To answer this kind of question, you can plot *networks* (where edges intersect) and calculate *free-flow travel time*, using the imputed maximum speed allowed on a specific highway type. In other words, when considering the maximum speed on a road, the time calculated to drive the distance is defined as *free-flow*. The `osmnx.speed` module calculates speed and travel times by providing kilometers per hour as `speed_kph` edge attributes. Complex networks are formed from the connections between nodes and edges. Analyzing these structures with the NetworkX Python package allows you to view a node as an element and the connection between nodes as an edge to study their relationship.

Let's try running a function to calculate travel time:

```
ox.speed.add_edge_speeds(G, hwy_speeds=None, fallback=None, precision=1)
ox.speed.add_edge_travel_times(G, precision=1)
```

Notice that you have four inputs here. Let's break these down and look at how the OSM documentation defines them:

G
: The input graph

hwy_speeds
: The mean speed value of all edges in highway type (if empty)

fallback
: A backup speed, in case the road type has no preexisting value

precision
: Rounding to the desired decimal precision

Only G, the NetworkX MultiGraph, is being passed into the function. If the origin (`orig`) and destination (`dest`) are both specified, you'll get a single list of nodes in a shortest path from the source to the target. These routes are distinct, to show how you can plot `travel_times` and `edge_speeds` to find the shortest routes between destinations.

You are setting weight to equal w instead of the default none. Here's what it looks like:

```
G = ox.add_edge_speeds(G)
G = ox.add_edge_travel_times(G)

w = 'travel_time'
orig, dest = list(G)[10], list(G)[-10]
route1 = nx.shortest_path(G, orig, dest, weight=w)
```

```
orig, dest = list(G)[0], list(G)[-1]
route2 = nx.shortest_path(G, orig, dest, weight=w)
orig, dest = list(G)[-100], list(G)[100]
route3 = nx.shortest_path(G, orig, dest, weight=w)

routes = [route1, route2, route3]
rc = ['r', 'y', 'c']
fig, ax = ox.plot_graph_routes(G, routes, route_colors=rc, route_linewidth=6,
figsize=(30, 30),node_size=0,bgcolor='#FFFFFF')
```

Figure 5-6 uses OSMnx's color options as well. *Color options* can be described as attribute values for each edge in a path. *Attribute values* can represent a property, such as data about how the graph was constructed, the color of the vertices when the graph is plotted, or simply the weights of the edges in a weighted graph. In Figure 5-6, the routes are represented by different colors.

Figure 5-6. Calculating the shortest distance between points in Los Angeles by computing weighted travel times

The full street network does consume resources, so let's select a specific neighborhood to get a more manageable visualization to query. In the code snippet that follows, I designated Culver City, California, to include all the private roads within the city limits (see Figure 5-7):

```
place_name = 'Culver City, California, United States'
G = ox.graph_from_place(place_name, clean_periphery=False,
network_type="all_private")
fig, ax = ox.plot_graph(G,figsize=(7,7),bgcolor='#FFFFFF',
    node_color="b",node_size=3)
```

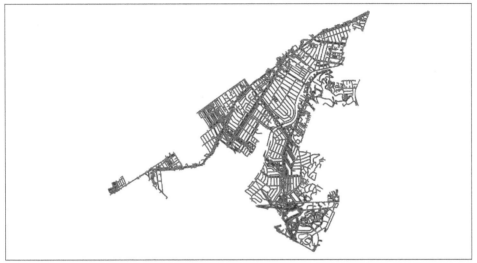

Figure 5-7. A neighborhood-level street network map of Culver City, California

If you are curious about colorizing nodes or edges, you can explore key:colour (*https://oreil.ly/zjof8*) options on the OSM wiki page or in the user reference documentation. Figure 5-8 shows the most common color codes but is certainly not exhaustive. The color variation is somewhat self-explanatory, but you can refer to the exact color on the OSM wiki (*https://oreil.ly/zjof8*). I find the chart useful to grab the color codes I use most frequently.

Sample	RGB hex	W3C colour name	Sample	RGB hex	W3C colour name
	#000000 (or #000)	black		#FFFFFF (or #FFF)	white
	#808080	gray (or grey)		#C0C0C0	silver
	#800000	maroon		#FF0000 (or #F00)	red
	#808000	olive		#FFFF00 (or #FF0)	yellow
	#008000	green		#00FF00 (or #0F0)	lime
	#008080	teal		#00FFFF (or #0FF)	aqua (or cyan)
	#000080	navy		#0000FF (or #00F)	blue
	#800080	purple		#FF00FF (or #F0F)	fuchsia (or magenta)

Figure 5-8. Key color options in OSM (not exhaustive, but a handy reference)

You have built a few street networks and maps in OSMnx. Now let's see what additional information you can extract.

Basic Statistical Measures in OSMnx

Visualizing street networks is meaningful, but the OSMnx framework also allows you to generate descriptive geometric and topological measures. Topographic measures can offer insights like the average number of nodes per street.

Circuity

To introduce the concept of circuity, let's calculate the average circuity in Manhattan and compare it to Staten Island. *Circuity* is the ratio of network distances to straight-line distances. Circuity is an important measure in urban networks; it indicates transportation efficiencies and may also show inequities between different geographic areas and populations. Circuity averages differ in different neighborhoods, on different subway routes, and with the length and time of travel. Researchers used circuity as part of a study that found a correlation between subway turnstile entries in New York City and COVID-19 cases and deaths (*https://oreil.ly/7qGjF*) between March and May 2020.

How circuitous your journey from point A to point B is, compared to other journeys, relies on a variety of factors, such as land-use systems, the design of transportation options, hubs, and cost. Network distances like circuity are calculated based on the actual grid and layout of a city, instead of "as the crow flies"—that is, a simple straight line between two points. You can always assume that actual distances to locations are quite a bit farther.

I will repeat the import details in case you are moving between different sections. There is a new line in the code:

```
%matplotlib inline
```

The % character indicates this is a Python "magic function." The argument is the line that follows:

```
ox.__version__
```

These *magic functions* (or dunder methods) with double underscores provide additional functionality. In this example, matplotlib will display the plotting command within the Jupyter Notebook right below the code.

Parentheses are not needed. This places the plotting output inside the Jupyter Notebook beneath the coding cell:

```
import osmnx as ox
import networkx as nx
import matplotlib.pyplot as plt #installs with osmnx
import pandas as pd
%matplotlib inline
ox.__version__
```

Here is a value for travel in Manhattan, which has a robust transportation framework:

```
# get the network for Manhattan, calculate its basic stats and show the average
circuity
stats = ox.basic_stats(ox.graph_from_place("Manhattan,New York,United States"))
stats["circuity_avg"]
```

The output is 1.0366712806100773.

You can now compare this value to other boroughs. Lower circuity values indicate higher efficiency. The other four boroughs of New York City present different transportation challenges. Traveling within Staten Island, for instance, is 4% less direct than traveling within Manhattan, as the output of the following code indicates:

```
# get the network for Staten Island, calculate its basic stats and show the
average circuity
stats = ox.basic_stats(ox.graph_from_place("Staten Island,New York,United States"))
stats["circuity_avg"]
```

This output is 1.0732034499369911.

Network Analysis: Circuity in Paris, France

Let's have a little fun and explore Paris, specifically, part of its urban infrastructure—the roadways. First, let's calculate circuity:

```
stats = ox.basic_stats(ox.graph_from_place("Paris, France"))
stats["circuity_avg"]
```

And the output: 1.0377408863844562.

Viewing the network statistics for a place is as simple as updating the place name (according to the Nominatim standards of OSM) and running the cell. The street network is then loaded, along with the area in square meters. Let's get Paris's street network and area:

```
# get street network, and area in square meters
place = "Paris, France"
gdf = ox.geocode_to_gdf(place)
area = ox.project_gdf(gdf).unary_union.area
G = ox.graph_from_place(place, network_type="drive")
```

Now let's calculate and merge the statistics:

```
# calculate basic stats, merge them, and display
stats = ox.basic_stats(G, area=area)
pd.Series(stats)
```

The counts of streets and networks are stored as nested dictionaries inside the statistics dictionary. The following code unpacks those nested dictionaries and then converts them to a pandas dataframe:

```
# unpack dicts into individual keys:values
stats = ox.basic_stats(G, area=area)
for k, count in stats["streets_per_node_counts"].items():
    stats["{}way_int_count".format(k)] = count
for k, proportion in stats["streets_per_node_proportions"].items():
    stats["{}way_int_prop".format(k)] = proportion
# delete the no longer needed dict elements
del stats["streets_per_node_counts"]
del stats["streets_per_node_proportions"]

# load as a pandas dataframe
pd.DataFrame(pd.Series(stats, name="value")).round(3)
```

This example is probably more granular than most of what you'll need, but I want you to see the data you can use to calculate measures such as shortest distance.[1]

At a glance, you can determine the number of streets emerging from each node. If you are interested in density, you can count the number of intersections or meters of linear street distance. Specifically, *linear street distance* represents a sum of total street length in an undirected representation of a network, meaning directionality (one-way and two-way streets) is not captured. That information is important for driving directions but not as important for walking directions.

---

1 If you would like the details behind these calculations and a case study, see Boeing, G. 2017. "OSMnx: New Methods for Acquiring, Constructing, Analyzing, and Visualizing Complex Street Networks." *Computers, Environment and Urban Systems*, 65: 126–139. *https://doi.org/10.1016/j.compenvurbsys.2017.05.004*. The user reference also provides a summary of the dictionary with attributes included.

Betweenness Centrality

Let's ask a data question: *how long is the average Parisian street?* Average street length is a nice linear proxy for block size. These measures provide urban-planning information about walkability and housing prices; for instance, a smaller block size translates to higher walkability and higher home prices. Figure 5-9 shows the overall output of our statistics.

Betweenness centrality measures how central a location or node is within a larger network or neighborhood. You can see that in Paris, the area with the highest betweenness centrality has 11% of all the shortest paths running through its boundaries. What does this tell us? Shorter road segments typically cluster around central business districts or historical districts or serve as conduits for navigating throughout a city.

	value
n	9684.000
m	18714.000
k_avg	3.865
edge_length_total	1817259.652
edge_length_avg	97.107
streets_per_node_avg	3.155
intersection_count	9142.000
street_length_total	1492943.191
street_segment_count	15172.000
street_length_avg	98.401
circuity_avg	1.023
self_loop_proportion	0.001
node_density_km	91.954
intersection_density_km	86.808
edge_density_km	17255.770
street_density_km	14176.227

Figure 5-9. Network statistics for the city of Paris, France

How important is the node? Think about a chain of command where information has to travel through a single person to be dispensed to the rest of the organization. If only 11% of the information has to pass through it, then in general, it isn't critical to the flow through the network. You can see the max node as a small red dot in the lefthand image (A) in Figure 5-10. In the following code cell, the shortest path between the nodes is calculated. The parameter weight considers the edge attribute length. The function will return the betweenness centrality:

```
# calculate betweenness with a digraph of G (ie, no parallel edges)
bc = nx.betweenness_centrality(ox.get_digraph(G), weight="length")
max_node, max_bc = max(bc.items(), key=lambda x: x[1])
max_node, max_bc
```

The output is (332476877, 0.1128095261389006).

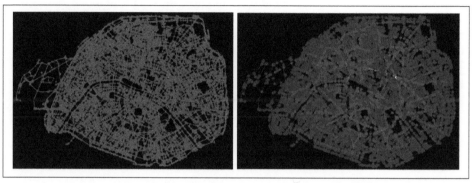

Figure 5-10. (A) Digraph (max node) of betweenness in Paris highlighting the maximum node of betweenness centrality; (B) Visualizing every node in the graph

Node color (nc) will retrieve a node with the highest centrality and observe the result in the graph object shown in Figure 5-10. Node size (ns) is calculated as well. You can adjust these values based on your desired visualization:

```
nc = ["r" if node == max_node else "grey" for node in G.nodes]
ns = [100 if node == max_node else 15 for node in G.nodes]
fig, ax = ox.plot_graph(G, node_size=ns, node_color=nc, node_zorder=5)
plt.show()
```

Next, try adding color to every node, to visualize 11% of all shortest paths relative to all nodes in Figure 5-10 (B):

```
# add the betweenness centrality values as new node attributes, then plot
nx.set_node_attributes(G, bc, "bc")
nc = ox.plot.get_node_colors_by_attr(G, "bc", cmap="plasma")
fig, ax = ox.plot_graph(
    G,
    node_color=nc,
    node_size=30,
    node_zorder=2,
    edge_linewidth=0.2,
    edge_color="w",
)
```

When observing street networks visually, it can be a challenge to identify important nodes and measures of centrality. Plotting the node attributes gives you an easily identifiable region for a deeper review. What other features are located nearby? How might this influence the values we observe at this node?

Network Types

How about we stick around in France a little longer? In Jupyter Notebooks, I often repeat the import functions for simplicity, to avoid scrolling up and down when running or rerunning a section. I will do the same thing here:

```
import osmnx as ox
import networkx as nx
import geopandas as gpd
import matplotlib.pyplot as plt
plt.style.use('default')
import pandas as pd
```

There are a few network types to explore. The most common are walk, bike, and drive, or you can even explore all. Some of those options can be quite congested in a city like Paris. Let's try drive and explore what that measures. Generate the figure in your Notebook:

```
place_name = "Paris, France"
graph = ox.graph_from_place(place_name, network_type='drive')
fig, ax = ox.plot_graph(graph)
```

You want edges (streets), so you will need to create a variable:

```
edges = ox.graph_to_gdfs(graph, nodes=False, edges=True)
```

What type of columns? What is the projection? The following code cells provide this information:

```
edges.columns
```

The output is:

```
Index(['osmid', 'name', 'highway', 'maxspeed', 'oneway', 'reversed', 'length',
       'lanes', 'geometry', 'junction', 'width', 'bridge', 'tunnel', 'access',
       'ref'],
      dtype='object')
```

You also will need to know the CRS to identify location in geographic space:

```
edges.crs
```

And the output:

```
<Geographic 2D CRS: EPSG:4326>
Name: WGS 84
Axis Info [ellipsoidal]:
- Lat[north]: Geodetic latitude (degree)
```

```
- Lon[east]: Geodetic longitude (degree)
Area of Use:
- name: World.
- bounds: (-180.0, -90.0, 180.0, 90.0)
Datum: World Geodetic System 1984 ensemble
- Ellipsoid: WGS 84
- Prime Meridian: Greenwich
```

I am a visual person, so I like to see the column headings with a sample of the data. Set that up next:

```
edges.head()
```

Now you can run the following code for a summary of the types of roads and how they are classified:

```
print(edges['highway'].value_counts())
```

Here is the output:

```
residential                           8901
primary                               2984
tertiary                              2548
secondary                             2542
unclassified                           648
living_street                          458
trunk_link                             194
trunk                                  146
primary_link                           118
[residential, living_street]            41
secondary_link                          37
[unclassified, residential]             30
tertiary_link                           20
motorway_link                           16
...
```

OSM contains information about roadways that you can access for a wide variety of uses. The infrastructure of any location or city can provide information about how easy it is to access transportation into and out of that location, which might influence the social nature of neighborhoods, communities, and larger networks of populations.

Customizing Your Neighborhood Maps

Let's look at how you can create, customize, automatically download, and compare street network maps. For this exercise, we'll travel back to Culver City and generate a neighborhood map.

Geometries from Place

The following code generates the map of Culver City, California, in Figure 5-11; including the tag building will add building footprints to the map:

```
place = "Culver City, California"
tags = {"building": True}
gdf = ox.geometries_from_place(place, tags={'building':True})
gdf.shape
```

I am defaulting to #FFFFFF (white) as a background color to make the map more easily printable:

```
fig, ax = ox.plot_geometries(gdf, figsize=(10, 10),bgcolor='#FFFFFF')
```

Go ahead and see what you can create. Figure 5-11 shows the city geometry with building footprints.

Figure 5-11. Using place to call a specific geometry by location in OSMnx for Culver City, California

Geometries from Address

Perhaps you have geographic coordinates for a location. You can explore that location by passing those coordinates to the geometries_from_address function (Figure 5-12). All of the geometries will transform to the CRS you select. For example, the projected coordinate system for the Netherlands is EPSG:28992.

Here is the code that generates the map in Figure 5-12:

```
gdf=ox.geometries.geometries_from_address((52.3716,4.9005),dist=15000,
tags={'natural':'water','highway':''})
gdf.to_crs(epsg=28992, inplace=True)
gdf.plot(figsize=(16,16))
gdf.plot(figsize=(16,16))
```

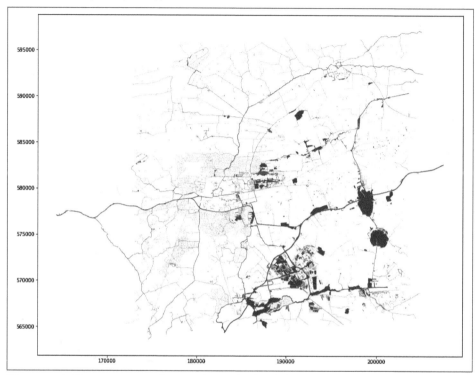

Figure 5-12. Viewing Amsterdam in the Netherlands by geographic coordinates and EPSG projection

You can also change the arguments to explore other geographies. Make sure to change the projection for the highest level of accuracy. (As a friendly reminder: the Earth is not a perfect sphere, so you need to select the best projection for the geographical area of interest to minimize distortion.) The following code includes additional tags:

```
gdf=ox.geometries.geometries_from_address(('Manhattan, NY'),dist=15000,
    tags={'natural':'water','building':'commercial','landuse':'commercial',
    'landuse':'residential','highway':'unclassified','highway':'primary'})
gdf.to_crs(epsg=2263,inplace=True)
gdf.plot(figsize=(20,29))
```

Figure 5-13 shows the output of this code. Thanks to the tags, this map shows water, commercial buildings, commercial and residential land use, and highways. These tags are explorative. See what happens with different combinations.

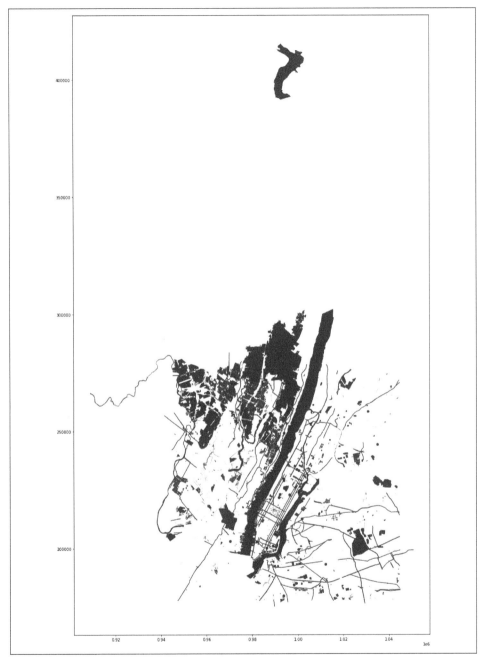

Figure 5-13. Using tags to add features to a map

In the next example, you'll use `custom_filter` to query specific ways to include in your graph, such as `motorway`. Specifying what you want with `custom_filter` will significantly reduce the resources needed to assemble the graph.

The terms `motorway_link` and `trunk_link` have specific meanings in OSM (as Open-StreetMap Wiki (*https://oreil.ly/DHjM8*) explains). They describe how highways are connected: on- and off-ramps, for example (see Figure 5-14).

Input the following code:

```
cf = '["highway"~"motorway|motorway_link|trunk|trunk_link"]'
G = ox.graph_from_place("Madrid", network_type="drive", custom_filter=cf)
fig, ax = ox.plot_graph(G, node_size=3,node_color='black',edge_linewidth=3,
bgcolor='#FFFFFF')
```

The `custom_filter` (cf) requires a `network_type` preset like `"highway"~"motorway"`. We want highways that are also motorways and the binary OR (|) operator in Python. The argument addresses both values, motorways and trunks (the ramps for entering and entering highways).

The output (Figure 5-14) shows a map of only motorways, this time in Madrid, Spain.

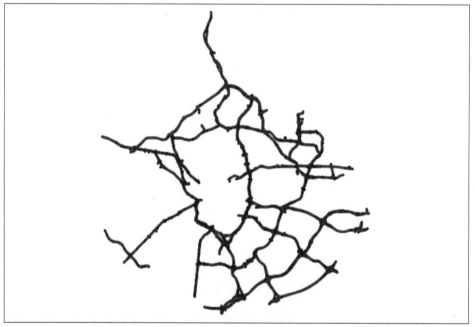

Figure 5-14. Motorways in Madrid using a custom filter (cf) to query only motorways

To appreciate the motorway map in context, you can readily bring up a more detailed map of Madrid with building footprints (Figure 5-15) using the following code:

```
place = "Madrid, Spain"
tags = {"building": True}
gdf = ox.geometries_from_place(place, tags)
fig, ax = ox.plot_footprints(gdf, figsize=(20, 20),alpha=1,color='black',
bgcolor='#FFFFFF',save=True,filepath='settings.imgs_folder/image.png',show=True,
dpi=600)
```

Notice that save is set to True. It is important to save your files. I am using the default filepath, but you can specify a precise folder or location, such as your Downloads folder.

Figure 5-15. Building footprints in Madrid, Spain

Figure 5-16 is an example of downloading the whole network (net work_type="drive") in the absence of a filter or customization. These renderings can be quite artistic and are often integrated into cartography and mapmaking or commercial products. Run this code to generate the map:

```
import osmnx as ox
import networkx as nx
import matplotlib.pyplot as plt
%matplotlib inline
place_name = 'Madrid,Spain'
G = ox.graph_from_place(place_name,network_type="drive")
fig, ax = ox.plot_graph(G,figsize=(20,20),bgcolor='#FFFFFF',
    node_color='black', node_size=0)
```

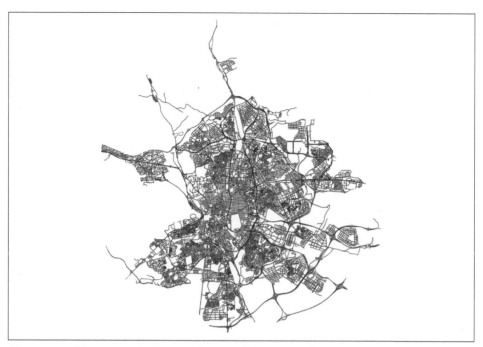

Figure 5-16. The networkx function brings all the networks into the graph

The last graph in this chapter is created with a different value for net
work_type="all". Figure 5-17 shows what happens to the output if you change
the geocode queries by swapping out the strings (all_private, all, bike, drive,
drive_service, walk):

```
G = ox.graph_from_place(
    "Madrid, Spain",network_type="all",
    retain_all=False,
    truncate_by_edge=True,
    simplify=True,

)

fig, ax = ox.plot_graph(G, figsize=(20, 20),node_size=0, edge_color="#111111",
edge_linewidth=0.7,bgcolor='#FFFFFF')
```

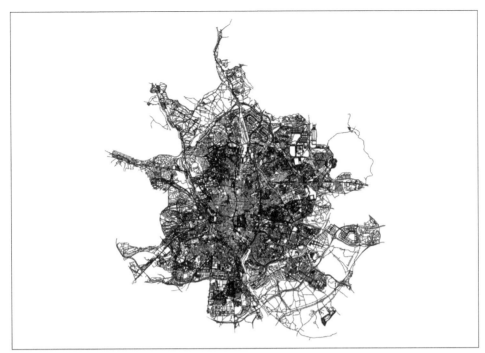

Figure 5-17. Madrid, Spain, observed in `network_type="all"` *in OSMnx*

You can experiment by choosing the other `network_type` options and observing how the output changes.

Working with QuickOSM in QGIS

You got acquainted with QGIS in Chapter 3, and I want to point out briefly that you can also use QGIS to interact with OSM data. QuickOSM in QGIS works similarly within a GUI. Open QGIS and select Vector in the toolbar, and you will see a "Quick query" option for Quick OSM (Figure 5-18). QuickOSM allows you to run these quick queries in the console and display results directly on a canvas.

Getting access to the level of data in GIS format within QGIS allows you to select a variety of amenities to include in a single map layer. This is an opportunity to explore `key:value` relationships outside of a Notebook as your data questions grow in scope.

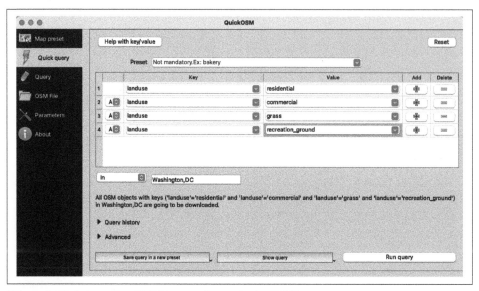

Figure 5-18. Integrating OSM key:value *pairs in QGIS*

Summary

In this chapter, you explored OpenStreetMap and learned how to generate street networks across the world as well as to use functions to calculate travel times, circuity, and other important measures.

OSM contains a wealth of open source geospatial data that you can access, model, project, visualize, and use to explore real-world street networks, points of interest, building locations, elevations, and more. I hope you will continue to play with it. Now I want to show you how to integrate this new skill with another tool, the ArcGIS Python API.

## CHAPTER 6

The ArcGIS Python API

The suite of client software and GIS developed by Esri, the global leader in GIS, is known as ArcGIS. It's an API and Python package in one that allows users to query for information hosted in ArcGIS Online or ArcGIS Enterprise. It's not open source, but Esri's leadership in the industry has produced quite a lot of free content and tutorials that you can access and explore. I will share a few learning resources, accessible tools, and information that you can access with the ArcGIS Python API. You'll need to use a Jupyter Notebook and ArcGIS Online to follow along.

Setup

There is a desktop application called ArcGIS Pro as well as a browser-based platform called ArcGIS Online; in this chapter, you'll be working with ArcGIS Online with the ArcGIS Python API.

Modules Available in the ArcGIS Python API

To extend the standard Python library, the ArcGIS Python API allows access to additional modules, available by dot notation. You can explore them in the documentation (*https://oreil.ly/FaGv3*), and I will provide more information as we use them in our coding. These modules include:

`Arcgis.gis`
    Connects you to ArcGIS Online

`Arcgis.features`
    Offers functions for working with groups of geographic elements

`Arcgis.geometry`
Allows input and output of different geometries (points or polygons, for example)

`Arcgis.geocoding`
Assigns location coordinates for map visualization

`Arcgis.geoenrichment`
Adds attributes to an area

I use open source or low-cost options when teaching geospatial skills to make this knowledge accessible to as many people as possible. Since most Esri users are professional or enterprise subscribers, few understand the cost burden for personal use. Whatever you choose, please read the documentation carefully before using it to avoid unintended costs.

The ArcGIS API is distributed as the package arcgis. In the following example, you'll install it via Conda, the suggested protocol, as you've done in several chapters in this book so far.

Installing ArcGIS Pro

Although ArcGIS Pro is only for Windows, the API lets non-Windows users access ArcGIS features without having to open up a Windows environment. I am going to assume that if you already have a full ArcGIS Pro license, you can follow the documentation to install Python Package Manager or Python Command Prompt (*https://oreil.ly/sXsRt*). To use the Pro license and work within your desktop environment with the API, you need to install ArcGIS on the same computer where you are running the commands.

If, like me, you have an ArcGIS Pro account but prefer to work in MacOS, you can use ArcGIS Pro offline. To do so, sign into your ArcGIS Pro account and, under Settings >> Licenses, check the box allowing you to work offline.

There are options to access publicly available resources even without a license, so let's explore those.

Setting Up Your Environment

Next, open your terminal. I suggest creating an environment for downloading Arc-GIS. (As you may recall, environments allow you to install compatible versions of packages.) Here, I am calling my environment `esriENV`.

Although it is possible to update individual packages within an environment, if I generate persistent errors when working in an environment, I delete the whole thing and re-create it from scratch. Here is a new Python geospatial package with commonly used packages that you may want to consider adding to any created space:

```
mamba install -c conda-forge geospatial
```

You may be curious about the Mamba install listed here. Since the writing of this book, mambaforge (*https://oreil.ly/1zzLj*) has emerged as a Conda-forge community project. Operationally, Mamba promises speed and compatibility with Conda packages. As you experiment, you'll reach your own preferred workflow. I have installed Mamba into my base environment and use it with my own projects, but the text has been written with the longer-term reliability of Conda environments.

Let's review how to create your environment and install the packages required for this Notebook. Enter the following code in the terminal, but instead of *myenv*, insert the name of *your* environment:

```
conda create --name myenv
```

To create an environment and include a specific version of Python to address package dependencies, enter the following code:

```
#conda create --name myenv python=x.x
conda create --name esriENV python=3.9
```

Now you must activate your new environment. We have only one channel or place to host installed packages at the moment, conda-forge, so we want all the dependencies that we need for installed packages to come from the conda-forge channel unless they exist only on default channels.

Alternatively, you may want to have the newest versions of packages on any channel in your list. If this is the case, use `conda config --set channel_priority strict`:

```
conda activate esriENV
#conda config --set channel_priority strict
```

Installing Packages

You are ready to install your packages:

```
conda install -c esri arcgis
```

Next, you will need to install the Jupyter Notebook; the IPython kernel, to execute the Python backend in Jupyter within a Conda environment; and Jupyter Notebook extensions to simplify coding tasks with tools like autocompletion:

```
conda install -c conda-forge notebook
conda install -c conda-forge nb_conda_kernels
conda install -c conda-forge jupyter_contrib_nbextensions
 conda install -c conda-forge bokeh
```

In your terminal, activate the ArcGIS environment using the name you gave it:

```
Conda activate esriENV
```

Once the packages are downloaded, you can once again open a Jupyter Notebook from your terminal so that you can document your work and evaluate the output step-by-step:

```
jupyter notebook
```

The Notebook will be installed along with the API and will open in a new window.

Connecting to the ArcGIS Python API

Now that you've finished installing, it's time to log in. First I'll show you how to log in anonymously; the section after that explains how to log in with a developer or individual account.

Connecting to ArcGIS Online as an Anonymous User

There are many ways to gain access to parts of ArcGIS for minimal or no charge; even without an ArcGIS account, you can use the resources, create maps, and share them, but with limited functionality. You can explore the options in Jupyter Notebook by running the code or exploring the documentation (*https://oreil.ly/kqmmF*). Another benefit of a public account is that it allows anonymous access.

You'll start by importing the Python API into your Jupyter Notebook. In this example, you'll be using a public account, so only free resources will be shown. No problem—there is a lot here for you to explore. To import GIS from the gis module, run:

```
from arcgis.gis import GIS

gis = GIS()
```

Connecting to an ArcGIS User Account with Credentials

If you have a user account, you can log in one of two ways: using the built-in login or using an API key. The advantage of logging in is that you can save any maps you create to your account.

Either way, you'll first need to import GIS using your terminal:

```
from arcgis.gis import GIS
```

To use the built-in login, run:

```
gis = GIS(username="someuser", password="secret1234")
```

To log in with an API key, use the following code (I've shortened the token):

```
gis = GIS(api_key="Request YOUR KEY AND INSERT IT HERE",
          referer="https")
```

When you connect with credentials and run the cell, the output will be your organi-zation's ArcGIS Online username. The username of my account is "datalchemy"; you will need to replace this with your own account credentials if you create an account:

```
gis = GIS(username="datalchemy", password="xxx")
gis
```

You can also log in with your password protected if you have an account and a login URL (listed in your organization's settings), again substituting your own credentials:

```
import arcgis
import getpass
username = input ("Username:")
password = getpass.getpass ("Password:")

gis = arcgis.gis.GIS("https://datalchemy.maps.arcgis.com", username, password)
```

Once you run the cell, a box will prompt you to enter your username and password.

Now you're ready to explore some layers of imagery.

Exploring Imagery Layers: Urban Heat Island Maps

Before we begin, a word about naming conventions. Figure 6-1 is a map of Chicago that I've named *map1*.

The designation is arbitrary: you may name your maps according to your preferen-ces, but the point is to avoid confusion when writing code that involves multiple maps. It is common to designate a map variable with an integer following it, or simply m. Select a method and be consistent. It should be distinct from the map() function. You can generate *map1* by running the following code in your Jupyter Notebook:

```
map1 =gis.map("Chicago,Illinois")
map1
```

This centers your map on Chicago, Illinois. You can set the area your map covers, called its *extent*, by zooming in on a specific location or by drawing a polygon.

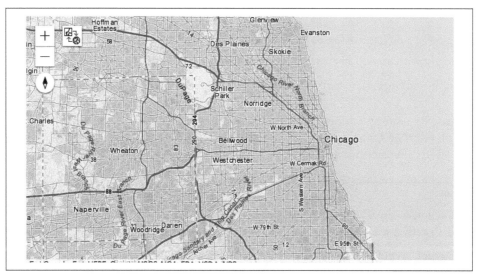

Figure 6-1. Map of Chicago

To access ArcGIS Online's help function, type:

```
gis?
```

If you type a dot and hit Tab, a dropdown menu will display additional properties that you can click for more information:

```
gis = GIS.
```

Now you can try layering some attributes onto your map. *Imagery layers* display data from image services. These resources allow you to apply rules to how images are displayed. You can search for them by identifying the item type as `Imagery Layer`.

You can use search terms in `gis.content.search()` to describe the features you would like to include in your map. Creating a variable assigned to `gis.content.search()` allows you to find available feature layers or other `item_type` data. Using the following code snippet, you can explore available Landsat 9 satellite views for your imagery layer. Different imagery layers have different properties. Here, I've restricted the search to the two options shown in Figure 6-2, which are the publicly available feature layers with the indicated terms:

```
from IPython.display import display

items = gis.content.search("Landsat 9 Views", item_type="Imagery Layer",
max_items=2)
for item in items:
    display(item)
```

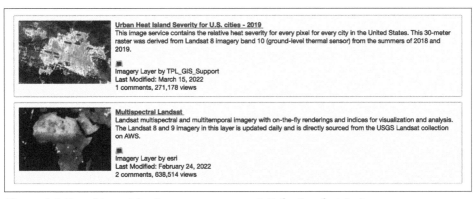

Figure 6-2. Searching with `gis.content.search()` *for Landsat 9 views*

`Item_type` in the code block includes web scenes, feature layers, geospatial data, basemaps, and interactive 3D environments with terrain. You can see more about items and item types in the documentation (*https://oreil.ly/MRi96*).

Web scenes let you visualize and analyze geospatial content containing all configuration settings, such as a basemap, styling, and extent. To explore these, you can substitute `web scene` for `item_type`.

We'll use web scenes to explore heat islands in Chicago. According to the US Environmental Protection Agency (*https://oreil.ly/3RWFQ*) (EPA), *heat islands* are:

> urbanized areas that experience higher temperatures than outlying areas. Structures such as buildings, roads, and other infrastructure absorb and re-emit the sun's heat more than natural landscapes such as forests and water bodies. Urban areas, where these structures are highly concentrated and greenery is limited, become "islands" of higher temperatures relative to outlying areas.

Urban heat island severity measures how city infrastructure absorbs and re-emits this heat.

To locate the image server URL and begin exploring heat islands on your map, click the title ("Urban Heat Islands") in your search results or pull up a list by calling `img_svc_url` in the next cell:

```
img_svc_url = 'https://server6.tplgis.org/arcgis6/rest/services/
heat_severity_2019/ImageServer'
```

This outputs:

```
[<Item title:"Urban Heat Island Severity for U.S. cities - 2019" type:
Imagery Layer owner:TPL_GIS_Support>,
 <Item title:"Multispectral Landsat" type:Imagery Layer owner:esri>]
```

You can try raster functions with this code, too. More about those in the next section, but for now, try this:

```
from arcgis.raster import ImageryLayer
landsat_urbanheat = ImageryLayer(img_svc_url)
landsat_urbanheat.properties.name
```

This outputs:

```
'Heat_severity_2019'
```

You have accessed the Urban Heat Island Severity imagery layer. Now add it to your map of Chicago:

```
map = gis.map('Chicago', zoomlevel=13)
map

map.add_layer(landsat_urbanheat)
```

The `landsat_urbanheat` layer is now added to your map, which should look similar to Figure 6-3.

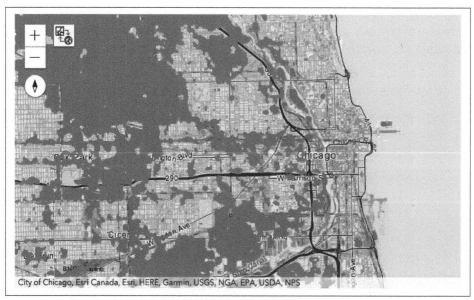

Figure 6-3. The map layer added to the map extent

Now you'll add the imagery layer from the URL for the Multispectral Landsat layer and assign it to the variable `landsat_ms`:

```
img_svc_url ='https://landsat2.arcgis.com/arcgis/rest/services/Landsat/MS/
    ImageServe'
from arcgis.raster import ImageryLayer
```

```
landsat_ms = ImageryLayer(img_svc_url)
landsat_ms.properties.name
```

The last line of code simply confirms the name of the layer.

This should generate the following output:

```
'Landsat/MS'
```

A description of the image layer is also available with a line of code:

```
landsat_ms.properties['description']
```

The output is detailed:

```
Multispectral Landsat image service covering the landmass of the World. This
service includes scenes from Landsat 8 and Global Land Survey (GLS) data from
epochs 1990, 2000, 2005 and 2010 at 30 meter resolution as well as GLS 1975
at 60 meter resolution. GLS datasets are created by the United States
Geological Survey (USGS) and the National Aeronautics and Space Administration
(NASA) using Landsat images. This service can be used for mapping and change
detection of urban growth, change of natural resources and comparing Landsat
8 imagery with GLS data.  Using on-the-fly processing, the raw DN values are
transformed to scaled (0 - 10000) apparent reflectance values and then
different service based renderings for band combinations and indices are applied.
The band names are in line with Landsat 8 bands; GLS data band names are mapped
along the same lines.
```

To explore additional raster function objects, visit the Common Data Types Documentation (*https://oreil.ly/MMRDP*). I also recommend keeping the ArcGIS Python API reference guide (*https://oreil.ly/DDDfa*) handy as you learn about raster functions and imagery layers.

Raster Functions

Raster functions process pixels from an image to one or more rasters without the need to download or create intermediate files for analysis. Access to raster functions is limited without additional credentials, but you still have the opportunity to explore Landsat multispectral imagery.

This time, we'll explore a map of Los Angeles, working our way through a few highlights:

```
from arcgis.gis import GIS
from arcgis.geocoding import geocode
from arcgis.raster.functions import *
from arcgis import geometry
import ipywidgets as widgets

import pandas as pd
```

You'll run another search in `gis.content.search()`. This time, you're looking for images outside of your organization (if you have created one) by indicating `outside_org=True`. This is the setting when you are accessing publicly available datasets:

```
landsat_item = gis.content.search("Landsat Multispectral tags:'Landsat on AWS',
'landsat 9', 'Multispectral', 'Multitemporal', 'imagery', 'temporal', 'MS'",
'Imagery Layer', outside_org=True)[0]
landsat = landsat_item.layers[0]
df = None
```

Your search results should include Multispectral Landsat, shown in Figure 6-4. To view the `landsat_item`, call it in a cell:

```
landsat_item
```

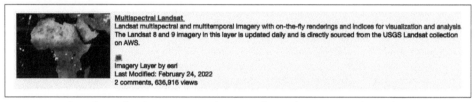

Figure 6-4. Search results showing a Multispectral Landsat imagery layer by Esri

For additional information, click on it or run the following code:

```
from IPython.display import HTML
HTML(landsat_item.description)
```

This map features *multispectral bands of light*, which are detected with instruments more sensitive than the human eye. As you learned in Chapter 1, multispectral bands capture image data within specific wavelength ranges across the electromagnetic spectrum to highlight different land-cover features. You can examine the name of each band, its minimum and maximum wavelengths, and other properties:

```
pd.DataFrame(landsat.key_properties()['BandProperties'])
```

The bands indicate details related to land mass, vegetation, and other types of land cover. Table 6-1 is an excerpt of the output, placed into a table for readability.

Table 6-1. Multispectral and Thermal Infrared Sensor Bands

Bands	Wavelength (micrometers)	Resolution (meters)
Band 1—Coastal aerosol	0.43–0.45	30
Band 2—Blue	0.45–0.51	30
Band 3—Green	0.53–0.59	30
Band 4—Red	0.64–0.67	30
Band 5—Near Infrared (NIR)	0.85–0.88	30

Bands	Wavelength (micrometers)	Resolution (meters)
Band 6—Short-Wave Infrared (SWIR) 1	1.57–1.65	30
Band 7—Short-Wave Infrared (SWIR) 2	2.11–2.29	30
Band 8—Panchromatic	0.50–0.68	15
Band 9—Cirrus	1.36–1.38	30
Band 10—Thermal Infrared (TIRS) 1	10.6–11.19	100
Band 11—Thermal Infrared (TIRS) 2	11.50–12.51	100

Create the map using your Jupyter Notebook:

```
landsat = landsat_item.layers[0]
landsat

m = gis.map('los angeles')
m
```

Now add the feature layer:

```
m.add_layer(landsat)
```

You can now view your map (Figure 6-5). You can also begin using raster functions.

Figure 6-5. A Los Angeles map created using Multispectral Landsat data

Here, you'll use a raster function to highlight color visualizations in your map. For example, color infrared shows bright red bands to indicate healthy vegetation, while natural color shows topography as we typically see it: green vegetation, blue water, brown soil.

Many of the options are available with a *dynamic range adjustment (DRA)* to enhance details and improve visibility when being matched to the range of your computer or other device. You can use the following code to generate a list of available raster functions, as shown in the output that follows:

```
for rasterfunc in landsat.properties.rasterFunctionInfos:
    print(rasterfunc.name)
```

This outputs a list:

```
Agriculture with DRA
Bathymetric with DRA
Color Infrared with DRA
Geology with DRA
Natural Color with DRA
Short-wave Infrared with DRA
Agriculture
Bathymetric
Color Infrared
Geology
Natural Color
Short-wave Infrared
NDVI Colorized
Normalized Difference Moisture Index Colorized
NDVI Raw
NBR Raw
None
```

Getting familiar with a few of these functions should pave the way for you to take a deeper look at other options in your independent explorations.

Let's try viewing `color_infrared`. You'll use the `apply` function:

```
From arcgis.raster.functions import apply
Color_infrared = apply (landsat, 'Color Infrared with DRA')
```

To visualize the map, call the `map` function:

```
m = gis.map('los angeles')
m.add_layer(color_infrared)
m
```

Healthy vegetation is now shown in bright red (Figure 6-6).

Figure 6-6. Los Angeles viewed with `color_infrared`

The best raster function for viewing vegetation is often the Normalized Difference Vegetation Index (NDVI). Figure 6-7 demonstrates the `NDVI_colorized` function, showing vegetation in green:

```
ndvi_colorized = apply(landsat, 'NDVI Colorized')
ndvi_colorized
```

Figure 6-7. Los Angeles NDVI

You have learned how to search for images and feature layers and how to apply raster functions to a specific location to highlight specific features. Next, we will explore attributes and take a look at the `arcgis.geometry` module.

Exploring Image Attributes

In this section, you'll explore the geometry of your Los Angeles map using the ArcGIS raster module get_samples (*https://oreil.ly/ffCTy*). This module lets you view sample point locations, spatial resolutions, and pixel values for a selected geometry. It's an example of the ArcGIS REST API (*https://oreil.ly/SWZTB*), which provides information about the architecture of web applications. In contrast to the general ArcGIS API documentation, which is focused primarily on access points, the ArcGIS REST API documentation (*https://oreil.ly/QdcRC*) provides the full scope of geospatial functions. It also describes the arguments (*https://oreil.ly/AXaKp*) that the get_samples operation can call:

```
get_samples(geometry, geometry_type=None, sample_distance=None, sample_count=None,
mosaic_rule=None, pixel_size=None, return_first_value_only=None,
interpolation=None, out_fields=None, slice_id=None)
```

Here you indicate the geometry and sample count of your map and return information about its image attributes, based on the extent you define—in this case, Los Angeles:

```
import arcgis
g = arcgis.geometry.Geometry(area['extent'])
```

Why geometry? When you create 3D models, data points like altitude and sun azimuth are useful for calculating hillshade. When working with the hillshade function, you are basically taking a 2D surface and rendering it as realistic 3D terrain.

Next, indicate the geometry and sample count:

```
samples = landsat.get_samples(g, sample_count=50,
                              out_fields='AcquisitionDate,OBJECTID,GroupName,
                              Category,SunAzimuth,SunElevation,CloudCover')
```

You can indicate an item to view in the samples list you have generated, or view all of them:

```
samples[10]
```

The output includes the location, an object ID, calculations of cloud cover, and pixel values, among other things:

```
{'location': {'x': -13150297.20625444,
  'y': 4059732.8562727477,
  'spatialReference': {'wkid': 102100, 'latestWkid': 3857}},
 'locationId': 10,
 'value': '1158 991 843 769 2850 1675 1030 44 21824 22410 22691',
 'rasterId': 3508198,
 'resolution': 30,
 'attributes': {'AcquisitionDate': 1647023302000,
  'OBJECTID': 3508198,
  'GroupName': 'LC09_L1TP_041036_20220311_20220311_02_T1_MTL',
```

```
 'Category': 1,
 'SunAzimuth': 144.81874084,
 'SunElevation': 45.83943939,
 'CloudCover': 0.0024},
'values': [1158.0,
991.0,
843.0,
769.0,
2850.0,
1675.0,
1030.0,
44.0,
21824.0,
22410.0,
22691.0]}
```

In Python, we use the `datetime` object for *time-series data*, or data collected at different time points. Again, you are sampling the `datetime` class in Python and can render the acquisition date by running the following code. (I'll provide additional details later in the chapter, but this is how we use the feature on samples data.)

Because Python accesses data based on zero-based indexing, using [0] requests the first index listed, or the first value in the sample:

```
import datetime
value = samples[0]['attributes']['AcquisitionDate']
datetime.datetime.fromtimestamp(value /1000).strftime("Acquisition Date: %d %b,
    %Y")
```

This outputs:

```
'Acquisition Date: 11 Mar, 2022'
```

By sampling values at this specific location, Los Angeles, you can estimate a spectral profile for specific points on your map.

```
m = gis.map('los angeles')
m

m.add_layer(landsat)
```

If you select a specific pixel, the spectral profile will plot all of the bands reflected at that location, as shown in Figure 6-8.

The next code lets you select a point on the canvas in your Jupyter Notebook to identify the spectral profile. Note that if you run this example from the ArcGIS Python API guide and request Landsat 9, it will generate an error because additional bands were added.

Figure 6-8. Map to click to generate the spectral profile in Figure 6-9

The `get_samples()` method will gather the pixel values contained in the sample data. The pixel values are *digital numbers*. To calculate them correctly, you'll need to convert them, first to floating-point numbers and then to integers. Digital numbers record the electromagnetic intensity of the pixel.

You installed the Bokeh package (*https://oreil.ly/QLQGJ*) into your Conda environment to allow interactive plots and charting. The Landset specification was based on Landsat 8 data and only included eight bands. This results in a Bokeh error. To fix this, run it with the new bands included:

```python
from bokeh.models import Range1d
from bokeh.plotting import figure, show, output_notebook
from IPython.display import clear_output
output_notebook()

def get_samples(mw, g):
    clear_output()
    m.draw(g)
    samples = landsat.get_samples(g, pixel_size=30)
    values = samples[0]['value']
    vals = [float(int(s)/100000) for s in values.split(' ')]

    x = ['1','2', '3', '4', '5', '6', '7', '8','9','10','11']
    y = vals
    p = figure(title="Spectral Profile", x_axis_label='Spectral Bands',
        y_axis_label='Data Values', width=600, height=300)
    p.line(x, y, legend_label="Selected Point", line_color="red", line_width=2)
    p.circle(x, y, line_color="red", fill_color="white", size=8)
    p.y_range=Range1d(0, 1.0)

    show(p)
```

```
print('Click anywhere on the map to plot the spectral profile for that location.')
m.on_click(get_samples)
```

You should get a message that Bokeh 2.4.2 (or your version number) has successfully loaded. Now return to the map and select your point. Click to generate the map shown in Figure 6-8 (yours will have different values, depending on the point you select).

The map you generated (Figure 6-8) is now interactive. Click anywhere on your new map to plot the spectral profile for that location. Now you can select different points and see what values you generate. Figure 6-9 shows the spectral profile of Band 9, used for detecting cirrus clouds, at the location I selected.

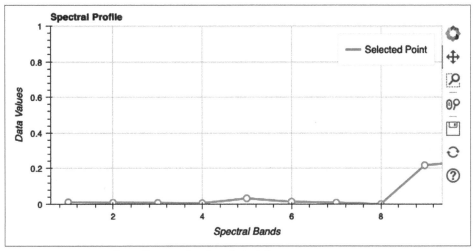

Figure 6-9. A spectral profile visualized with Bokeh

Depending on the unique location you select when clicking on the Landsat 9 imagery layer, your spectral profile will be different. Refer to Table 6-1 to identify the bands included. These represent a few applications of the `get_samples` method in the ArcGIS Python API.

Improving Images

Raster functions allow you to do things like extract specific bands to examine land use, vegetation, or fire; continuous data such as temperature; scanned images; and satellite images, among other things. A *stretch function* allows you to adjust the brightness and contrast on your map. The following code selects bands 3, 2, and 1 (red, green, and blue) from the visible spectrum and imports the `stretch raster.function`:

```
from arcgis.raster.functions import stretch, extract_band
naturalcolor = stretch(extract_band(landsat, [3,2,1]),
```

```
             stretch_type='percentclip', min_percent=0.1, max_percent=0.1,
                          gamma=[1, 1, 1], dra=True)
```

```
naturalcolor
```

The percent clip minimum (percentclip) that this code sets will exclude the lowest
10% of values from the stretch to apply to the raster. This is useful if most of the
pixels are within a specific range. Setting the stretch_type trims away the outliers to
redistribute the histogram of values, resulting in the image in Figure 6-10.

Figure 6-10. Natural color bands depicting what the eye would see without enhancement

The image in Figure 6-10 is what the eye would see without any enhancement.
Depending on the features you are looking for, there may be other bands to explore.
We will look at other bands in the following section.

Comparing a Location over Multiple Points in Time

The ArcGIS API allows you to compare images of the same location over different
points in time using a map widget called a *time slider*, which enables you to animate a
map using configurable properties like start and end times and intervals in between.
For this example, I've selected a city and a satellite basemap.

The map.zoom widget accepts values between 0 and 23. Level 9 gives a panned-out
global view; level 10 is somewhere between a large metropolitan area and a city; and
level 20 is the level of individual buildings. I have never zoomed to 23, but doing
so would result in the most detailed possible view. There are limitations to how
much you can zoom, though, depending on the scale of the visualization parameters
(conversion rates are available in the documentation (*https://oreil.ly/mbtnA*)). You
can change the zoom level and process the data, but the resolution will stay the same.
For example, all Landsat imagery has a resolution of 15 meters. This means that
the satellite image captures details on the ground that are 15 meters or larger. (For
comparison, the Hollywood sign in Los Angeles and a standard semitruck trailer are
both about 15 meters long.)

Try zooming to level 10:

```
map = gis.map("los angeles")
map.basemap = "satellite"
map.zoom = 10
map
```

The name of the `item title` is printed as output for reference:

```
landsat_item = gis.content.search("Landsat Multispectral tags:'Landsat on AWS',
'landsat 9', 'Multispectral', 'Multitemporal', 'imagery', 'temporal', 'MS'",
'Imagery Layer', outside_org=True)[0]
print(landsat_item)
<Item title:"Multispectral Landsat" type:Imagery Layer owner:esri>
```

You have searched for a Landsat item, and the recovered item is confirmed by its title in the output. Different sublayers will have different details provided.

Add the layer to the map:

```
map.add_layer(landsat_item)
```

Request confirmation that the `map.time_slider` is available:

```
map.time_slider
```

It should return the output `True`.

You can adjust the dates to a specific interval. The following code measures 10-day intervals between the selected start and end times:

```
from datetime import datetime
map.time_slider = True
map.set_time_extent(start_time=datetime(2021, 12, 12), end_time=datetime(2022, 4,
12), interval=10, unit='days')
```

This code outputs the map in Figure 6-11, which includes the time-slider widget.

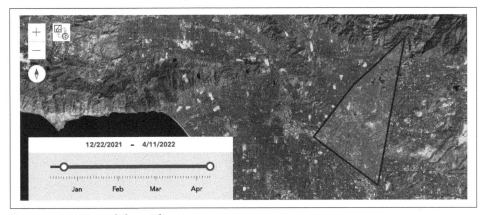

Figure 6-11. Time-slider widget

Select the `map.draw(polygon)` option in the code cells, and then draw a polygon with your cursor. If you have a specific location of interest, enter its coordinates:

```
from arcgis.geometry import Point

pt = Point({"x" : 34.092809, "y" : -118.328661,
            "spatialReference" : {"wkid" : 3309}})
```

A *spatial reference* or *well-known ID (WKID)* defines tolerance and resolution for the location you have selected (see the Spatial Reference List (*https://oreil.ly/AecRp*)).

Options for `map.draw` include selecting a point (`pt`), `polyline`, or `polygon`:

```
map.draw(pt)
map.draw('polyline')
map.draw('polygon')
```

When you select the polygon and run the code cell, simply return to the map and you will be able to trace a polygon. The time-slider widget allows comparison across a time series for an area of interest. When working with layers, you will often want to filter what appears on your canvas.

Filtering Layers

You may want to filter your layers: for instance, perhaps you only want to see Landsat layers from a specific collection (*https://oreil.ly/iMcfn*) with less than 10% cloud cover. For this, it's helpful to get familiar with Landsat's data dictionaries (*https://oreil.ly/9fR5U*) for complete information about data elements, attributes, and names within a database or system.

In the following code snippet, `WRS_Row` provides information about the orbital path of the satellite that took the image:

```
selected = landsat.filter_by(where="(Category = 1) AND (CloudCover <=0.10) AND
    (WRS_Row = 36)",
                geometry=arcgis.geometry.filters.intersects(area['extent']))
```

The value 36 indicates the northern hemisphere. (You can find these details in the Landsat data dictionary, the collection background, or elsewhere in Landsat's information.)

Information you can access is shown in the HTML item description you generated earlier. Now you can view the table you just created in a dataframe (`df`):

```
fs = selected.query(out_fields="AcquisitionDate, GroupName, Best, CloudCover,
    WRS_Row, Month, Name",
            return_geometry=True,
            return_distinct_values=False,
            order_by_fields="AcquisitionDate")
```

Because you are comparing dates, you'll want to see the oldest acquisition date as well as the most recent. To view your output clearly, run it in the code cell. The important takeaway is to see the column headings and where to find identifying information (Figure 6-12). You can view the first five rows by calling `df.head()` as follows:

```
df = fs.sdf
df.head()
```

	OBJECTID	AcquisitionDate	GroupName	Best	CloudCover	WRS_Row	Month	Name	SHAPE
0	3106827	1977-05-30	p043r036_2x19770530	95957036	-0.01	36	5	p043r036_2dm19770530_z11_MS	{"rings": [[[-12889114.8652, 4207339.365500003...
1	3106857	1979-06-08	p044r036_2x19790608	95956036	-0.01	36	6	p044r036_2dm19790608_z11_MS	{"rings": [[[-13049001.6684, 4177798.783399999...
2	3098905	1989-06-28	p041r036_5x19890628	91959036	-0.01	36	6	p041r036_5dt19890628_z11_MS	{"rings": [[[-13094672.6074, 4002345.807499997...
3	3098872	1990-05-07	p040r036_5x19900507	91960036	-0.01	36	5	p040r036_5dt19900507_z11_MS	{"rings": [[[-12902153.7239, 4064989.857699997...
4	3089612	2000-04-24	p040r036_7x20000424	88960036	0.00	36	4	p040r036_7dt20000424_z11_MS	{"rings": [[[-12872961.873599999, 4197538.7916...

Figure 6-12. Acquisition date output

Now run `df.tail()` and select a recent acquisition date. You can run the code in your Jupyter Notebook:

```
df = fs.sdf
df.tail()
```

The following command gives the shape of the data:

```
df.shape
```

The output informs us that the data has 9 columns and 193 rows: `(193,9)`.

If you only want the acquisition date, you can run this data, or any of the other columns, from the dataframe:

```
df['Time'] = pd.to_datetime(df['AcquisitionDate'], unit='ms')
df['Time'].head(10)
```

This outputs:

```
0    1977-05-30
1    1979-06-08
2    1989-06-28
3    1990-05-07
4    2000-04-24
5    2000-05-01
6    2005-05-15
7    2005-05-24
8    2009-04-16
9    2009-05-11
Name: Time, dtype: datetime64[ns]
```

Bands have specific wavelengths that are used in statistical calculations. You want to avoid overlapping pixels in an imagery layer because this can distort the calculations.

You can use the default method, where the pixel value is calculated from the last dataset (Figure 6-13):

```
m3 = gis.map('los angeles', 7)
display(m3)
m3.add_layer(selected.last())
```

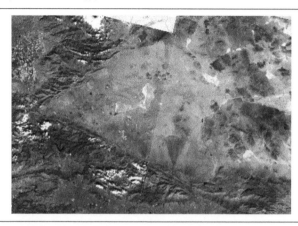

Figure 6-13. Overlapping pixels using the last() *method*

Alternatively, you can request the first() method and calculate pixel value from the first raster dataset (output shown in Figure 6-14):

```
m3 = gis.map('los angeles', 7)
display(m3)
m3.add_layer(selected.first())
```

Figure 6-14. Overlapping pixels using the first() *method*

Querying the data by OBJECTID, aided by the df(head) and df(tail) data, you can compare images captured at different times:

```
old = landsat.filter_by('OBJECTID=3106827')

new = landsat.filter_by('OBJECTID=3558253')

from arcgis.raster.functions import *
```

The stretch function (*https://oreil.ly/AsmeR*) improves visibility by spreading the pixel values. There are different types of stretch, but here you are simply removing any extreme values by calculating the standard deviation (stddev):

```
diff = stretch(composite_band([ndvi(old, '5 4'),
                               ndvi(new, '5 4'),
                               ndvi(old, '5 4')]),
                               stretch_type='stddev', num_stddev=2.5, min=0,
                               max=255, dra=True, astype='u8')
diff
```

This will output Figure 6-15. The green swaths indicate increases in vegetation density, while magenta shows a decrease within the timeframe selected by OBJECTID.

Figure 6-15. ArcGIS functions showing difference in images over time

Perhaps there is a certain threshold you want to capture. Try measuring only areas where the threshold change is above 10%. Use the following code:

```
threshold_val = 0.1
masked = colormap(remap(ndvi_diff,
                        input_ranges=[threshold_val, 1],
                        output_values=[1],
                        no_data_ranges=[-1, threshold_val], astype='u8'),
                  colormap=[[1, 124, 252, 0]], astype='u8')

Image(masked.export_image(bbox=area['extent'], size=[1200,450], f='image'))
```

The output (Figure 6-16) renders those areas in green.

Figure 6-16. Masked threshold value for changes in vegetation index (NDVI)

Let's render our map once again:

```
m = gis.map('los angeles')
m
```

The final image is displayed in Figure 6-17. You can now see the masked threshold by adding the layers to the map:

```
m.add_layer(diff)
m.add_layer(masked)
```

Figure 6-17. The combined image at the requested threshold displaying the masked threshold

Summary

Now that you've customized and explored a few publicly available resources from ArcGIS API for Python, you should have some familiarity with raster functions and image layers. To continue learning, try working with these functions and layers independently and see what other queries you can access.

## CHAPTER 7

GeoPandas and Spatial Statistics

Maps are beautiful. The stories they tell can be so captivating that it's easy to unintentionally disregard the geospatial statistics within them. But geospatial maps are not simply static images. There is information embedded in them, such as attributes associated with a specific feature in a GIS layer or pixel densities observed in raster images.

Python has a variety of packages for working with geospatial data. If you are familiar with Python, you likely know pandas (*https://oreil.ly/j6YBd*), a data analysis tool built specifically for Python. Pandas allows us to read a wide variety of data types into a dataframe: a set of tables containing rows (which denote records) and columns (denoting attributes). GeoPandas (*https://oreil.ly/pb8lC*) is an extension of pandas that lets you manipulate geometric and geospatial data using what it calls a *GeoDataFrame*: a geospatial dataframe in which each row is a spatial feature, such as a point, line, or polygon.

This chapter will show you how to analyze your data and create maps using Geo-Pandas and GeoDataFrames, as well as some other important packages, such as matplotlib (*https://matplotlib.org*) to visualize data and Census Data API. You will also learn how to access geospatial files and delve deeper into demographic data by creating and comparing demographic maps.

Installing GeoPandas

To install GeoPandas, you'll use `conda` and `conda-forge` in your terminal. As in previous chapters, I'll also show you how to create an environment for all the files you'll need to work with the data in this chapter.

Installing GeoPandas can be a little tricky. Think of it as a moody teenager. It often wants to go first (importing) and likes the latest trends (pay attention to versioning of dependencies). It also prefers a catchy nickname: typically, GeoPandas is called gpd, but from time to time, the code will write out geopandas instead. Pay attention and adjust your variables.

To set up your environment, start with this:

```
conda create --name geop_env python = 3.X
```

You can add your Python version here.

```
MacBook-Pro-8:~ USERNAME$ conda activate geop_env
```

Now you can start installing files into the geop_env:

```
conda install -c conda-forge geopandas
```

Now you can add any packages you would like to access to your geop_env. If you would like a collection of packages for geospatial analysis, I suggest downloading the geospatial package, but for now you can add packages individually if you prefer:

```
conda install -c conda-forge geospatial
```

GeoPandas simplifies working with geospatial data in Python. Let's explore the expanded operations that enable you to perform spatial operations on geometric data.

Working with GeoJSON files

GeoPandas lets you work with GeoJSON files. You may have heard of JSON, a common data exchange format. GeoJSON is a data exchange format based on JSON that's specifically designed to represent geographic data structures and their nonspatial attributes. Working with GeoJSON files makes it fairly simple to identify coordinates and locations for your maps.

Figure 7-1 shows a map from *geojson.io* (*https://oreil.ly/JmXP4*). Geojson.io is an open source tool that simplifies creating, viewing, and sharing your maps. You can perform a variety of functions on the internet without downloading any data to your computer.

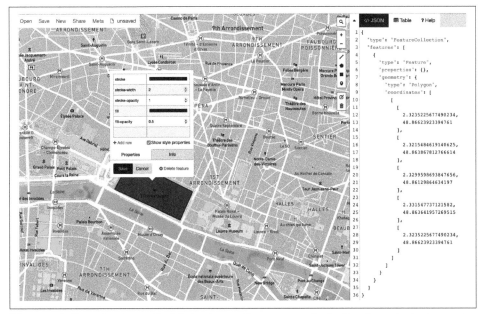

Figure 7-1. Zooming in on Paris's 1st Arrondissement and selecting boundaries to create a GeoJSON file

Click on the link *geojson.io* and zoom into a country. I chose France because, well, it's France. An easy way to locate coordinates for a specific location is to select the region you are interested in and drag the polygon around the area. The tools for selecting a polygon are visible on the right edge of the mapping dashboard. Try locating Paris, France, on the map. Find the Tuileries Garden and draw a polygon around it. Once you save the polygon, you can view formatting options and style properties by clicking within the boundary.

The Meta tab in the top menu (shown in Figure 7-1) allows you to generate a *bounding box (bbox)*, which contains the coordinates and set of points you are interested in mapping. It defines the spatial location of an object of interest and its coordinates. You can also load a string in the well-known text (WKT) markup language for vector geometry objects.

The buttons in the lower left corner, visible in Figure 7-2, let you toggle between Mapbox (vector tiles), Satellite (raster data), and OSM (OpenStreetMap, which you learned about in Chapter 5).

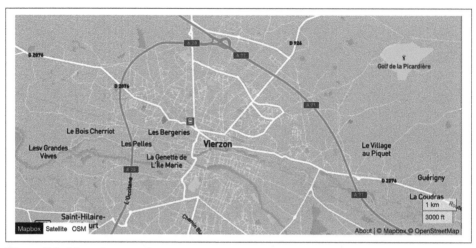

Figure 7-2. Map viewer options

If you want to save this file, you can do so in several different formats, including .csv, .txt, and a shapefile. Then you can import it to QGIS, ArcGIS, or a platform of your choice for additional exploration and analysis. Spreadsheet data is stored as .csv or .txt and often contains location information, such as coordinate data for longitude and latitude, addresses, or zip codes. Shapefiles are Esri vector files that capture location, shape, and related files about locations.

Creating a GeoDataFrame

A GeoDataFrame is much the same as a pandas dataframe but with an extra column of geometric data. Many types of vector data describe discrete data with a fixed location while continuous data is often represented by raster data (although both can be either).

GeoPandas makes a few dataset examples available for exploration. You are going to access an example from NYC Boroughs, abbreviated as nybb:

```
geopandas.datasets.available = ['naturalearth_cities', 'naturalearth_lowres',
'nybb']
```

These files represent the positions of cities, the contours of different countries, and the boundaries of New York City's boroughs. To access them, you will need to include the name of the dataset as a parameter:

```
geopandas.datasets.get_path("nybb")
```

You can open this in a Jupyter Notebook from your terminal:

```
(geop_env) MacBook-Pro-8:~ USERNAME$ jupyter notebook
```

Now, with the notebook open, you can create a dataframe. First, import the packages you'll need into your environment:

```
%matplotlib inline

import requests
import pandas as pd
import geopandas as gpd

from scipy.spatial.distance import cdist
```

Next, you'll need to retrieve the dataset and call for the active geometry column (in this instance, "geometry") by creating a dataframe called world.geometry.name. When you call head(), the top five rows and the columns in the dataset will be returned:

```
boros_world = gpd.read_file(gpd.datasets.get_path('nybb'))
print(f"{type(boros_world)}, {boros_world.geometry.name}")
print(boros_world.head())
print(boros_world.geometry.geom_type.value_counts())
```

You should get the following output:

```
<class 'geopandas.geodataframe.GeoDataFrame'>, geometry
  BoroCode  BoroName  Shape_Leng Shape_Area \
0    5    Staten Island 330470.010332 1.623820e+09
1    4    Queens 896344.047763 3.045213e+09
2    3    Brooklyn 741080.523166 1.937479e+09
3    1    Manhattan 359299.096471 6.364715e+08
4    2    Bronx 464392.991824 1.186925e+09

            geometry
0 MULTIPOLYGON (((970217.022 145643.332, 970227....
1 MULTIPOLYGON (((1029606.077 156073.814, 102957...
2 MULTIPOLYGON (((1021176.479 151374.797, 102100...
3 MULTIPOLYGON (((981219.056 188655.316, 980940....
4 MULTIPOLYGON (((1012821.806 229228.265, 101278...
MultiPolygon 5
dtype: int64
```

Make note of the index assignments in the first column. You will be able to isolate a single borough by referring to it by its index location. BoroCode and BoroName are additional identifiers you can specify, but you may notice the simplicity of calling by index in the code snippets that follow.

You are now working with a GeoDataFrame. You can use it to generate the plot in Figure 7-3 by running the following code in your Jupyter Notebook:

```
boros = gpd.read_file(gpd.datasets.get_path('nybb'))
ax = boros.plot(figsize=(10, 10), alpha=0.5, edgecolor='k')
```

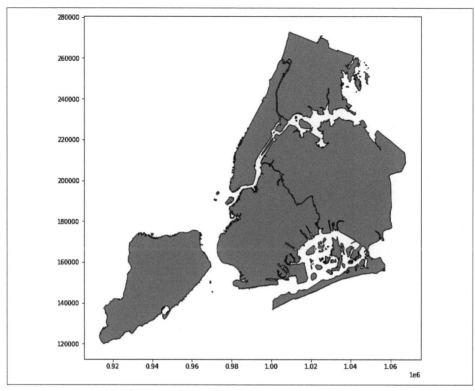

Figure 7-3. Locations of New York City boroughs in a GeoDataFrame

You can change the size of the figure using the `figsize` argument. In the example in Figure 7-3, an alpha value of less than 1 adjusts the transparency. Edge color can be customized as well. You can find details in the matplotlib documentation (*https:// oreil.ly/jy6fs*).

I prefer maps without the visible axes in most instances. To remove the frames, simply add:

```
ax.set_axis_off()
```

You will next explore the data by interactive map, so the area coordinates are not necessary. That will generate the image in Figure 7-4.

Figure 7-4. Locations of New York City boroughs without the axis frame

You can request a single borough by calling out a row number. Staten Island is in row 0 of the GeoDataFrame you generated a few pages ago. You can access a location by its index number using the callable function `DataFrame.loc` (*https://oreil.ly/vSEXO*). I think of `loc` as an abbreviation for "label of column" since it requires an input label, such as "index" or "geometry." The abbreviation `iloc` is for integer location index/integer of column. Substitute the name of the dataframe you created:

```
boros.loc[0,'geometry']
```

This outputs the image of Staten Island shown in Figure 7-5.

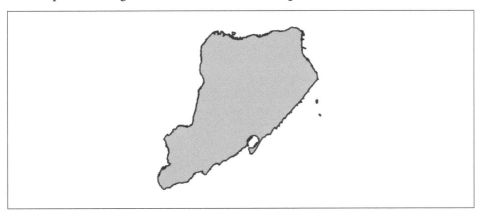

Figure 7-5. Calling a single borough using the `.loc` *function*

You can see the boundaries of Staten Island borough in Figure 7-5, but it would be helpful to have more information now that you know how to access the GeoData-Frame. You can also find the area of a borough:

```
gdf = gdf.set_index("BoroName")
gdf["area"] = gdf.area
gdf["area"]
```

This outputs:

```
BoroName
Staten Island    1.623822e+09
Queens           3.045214e+09
Brooklyn         1.937478e+09
Manhattan        6.364712e+08
Bronx            1.186926e+09
Name: area, dtype: float64
```

To view a borough's boundary, run:

```
gdf['boundary'] = gdf.boundary
gdf['boundary']
```

This outputs:

```
BoroName
Staten Island    MULTILINESTRING ((970217.022 145643.332, 97022...
Queens           MULTILINESTRING ((1029606.077 156073.814, 1029...
Brooklyn         MULTILINESTRING ((1021176.479 151374.797, 1021...
Manhattan        MULTILINESTRING ((981219.056 188655.316, 98094...
Bronx            MULTILINESTRING ((1012821.806 229228.265, 1012...
Name: boundary, dtype: geometry
```

In the interactive map (Figure 7-6), you can retrieve information by hovering over the borough. The reference distance is 0 because Staten Island's index number is 0. Clicking on any of the other boroughs will tell you the distance from Staten Island. The gdf.explore function generates an interactive map:

```
gdf.explore("area", legend=True)
```

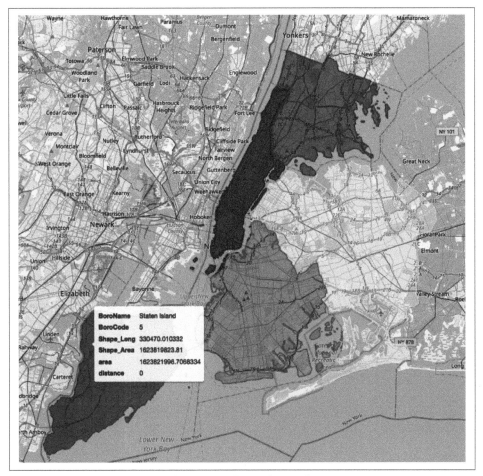

Figure 7-6. Interactive map of the New York City boroughs

Adding geospatial capabilities like geometry to pandas dataframes is only the beginning. Census data is a treasure trove of open source data, and knowing how to use GeoPandas will help you expand your ability to access and ask questions of this important dataset. Let's look at some more census data.

Working with US Census Data: Los Angeles Population Density Map

In this exercise, you'll calculate population density by census tract across the county of Los Angeles, California.

Be careful when working with Python census packages. If a developer loses interest in maintaining them or is overwhelmed with other tasks, your maps and graphics will no longer function. I learned that the hard way while writing this chapter. The good news is that it reminded me of the fail-safe basics that will always be reliable.

If you need a high-quality map of a census data question for professional services or research, you can build one in an open source platform like QGIS or by using proprietary software like ArcGIS. This example will walk you through the basics. When you understand how the Census API is structured and are able to write code outside of a fancy Python package, access to one of the most powerful publicly available datasets is at your fingertips.

The first step is locating the data on *census.gov* (*https://oreil.ly/k6wuo*). Data from the 2020 decennial census is becoming available for analysis depending on your area of interest. I rely on Census Reporter (*https://oreil.ly/TnwCF*) for identifying and downloading datasets. Most repositories have the ACS data (often called the "long-form census"), which is collected each year from a proportion of the nation and bundled yearly or in five-year vintages. The API fetches the last five-year ACS vintage by default. This is an easy-to-navigate first stop to holistically understand the type of data captured.

Census Reporter has ready-to-use neighborhood data for many metropolitan areas, and that is the data you will see in these examples.

 For more advanced users, IPUMS (*https://www.ipums.org*) (formerly the Integrated Public Use Microdata Series, now known by its acronym) offers global harmonized datasets from census and survey data with supporting documentation. You can use this to integrate ACS and decennial census data with merged data from other sources.

Accessing Tract and Population Data Through the Census API and FTP

Accessing the Census API, through the FTP site or directly, is the foundation for being able to work outside the Census GUI. This will help you address bigger questions, with data ready for analysis in notebooks or QGIS.

FTP sites provide quick and easy access to large files that can be challenging to navigate. If you are working in a MacOS environment, you may need to adjust a few settings, but I prefer the reliability and lack of clutter of accessing census data through FTP.

Select System Preferences >> Sharing and check Remote Login. Select which users to allow access.

Next, navigate to your Finder window and select Go. In the dropdown menu, you will see a Connect to Server option (Figure 7-7).

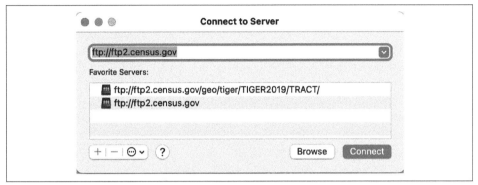

Figure 7-7. Connecting to the US Census Bureau FTP site

The FTP site (*ftp://ftp2.census.gov/geo/tiger/*) is accessible through the link. You will be able to scroll through the available files. You can also gain access through the server site (*https://oreil.ly/OlwXw*). The summary files (*https://oreil.ly/bsd2s*) are where most of the files of interest are located.

You are going to explore decennial population data from the US census. The geographical data included here is from the 2019 ACS one-year survey. The census tables for population are from the 2010 census.[1] Google Colab (*https://oreil.ly/ytceM*) works well for this exercise because the file structure is readily available, and once your session is over, the files are removed.

You will need to download the following packages not already included in Colab:

```
!pip install geopandas #install separately to avoid errors

import requests # accessing from the CensusAPI
import sys  # checking python version
import pandas as pd # working with dataframes
import geopandas as gpd # creating maps
import folium as fm # creating HTML map complete with OSM basemap
```

I learned an important lesson when working with census data: always know the version of your packages, in case there are dependencies (or conflicts!). This is a helpful practice to add to your notebooks. Use the following code to display versions:

---

1 The final data for the 2020 census was not yet available as of this writing. When it becomes available, you will only need to replace the link in the code cells with an updated link, and you will have the updated information.

```
# Display package versions
print("Python Version", sys.version)
print("requests version:", requests.__version__)
print("pandas version:", pd.__version__)
print("geopandas version:", gpd.__version__)
print("folium version:", fm.__version__)
```

Now you can request census tract information from the FTP server. You are looking for shapefiles for the census tracts. These are *.zip* files; note that the files you need may be bundled into a larger file. Download the topologically integrated geographic encoding and referencing (TIGER) files (*https://oreil.ly/THPPT*) for California: FIPS 06. (Federal Information Processing Standards Publications, or FIPS, files designate state and county equivalents.) TIGER shapefiles contain geographic information but not demographics, so we will add that information shortly. The link identifies the US Census Bureau FTP site and denotes geographic files from 2019 at the tract level in California:

```
'ftp://ftp2.census.gov/geo/tiger/TIGER2019/TRACT/tl_2019_06_tract.zip'
```

Once you identify the file in the FTP, to retrieve it, you need the !wget call. The wget utility retrieves FTP files from web servers:

```
!wget ftp://ftp2.census.gov/geo/tiger/TIGER2019/TRACT/tl_2019_06_tract.zip
```

The quotations on the *.zip* file path are not needed for the !wget call, and the file will unzip into your Colab notebook. You can watch the progress right in the notebook as it unzips to 100%. Your output should look like this:

```
--2022-08-22 15:52:00--  ftp://ftp2.census.gov/geo/tiger/TIGER2019/TRACT/
tl_2019_06_tract.zip
           => 'tl_2019_06_tract.zip'
Resolving ftp2.census.gov (ftp2.census.gov)... 148.129.75.35, 2610:20:2010:a09:
1000:0:9481:4b23
Connecting to ftp2.census.gov (ftp2.census.gov)|148.129.75.35|:21... connected.
Logging in as anonymous ... Logged in!
==> SYST ... done.    ==> PWD ... done.
==> TYPE I ... done.  ==> CWD (1) /geo/tiger/TIGER2019/TRACT ... done.
==> SIZE tl_2019_06_tract.zip ... 29388806
==> PASV ... done.    ==> RETR tl_2019_06_tract.zip ... done.
Length: 29388806 (28M) (unauthoritative)

tl_2019_06_tract.zi 100%[===================>]  28.03M   262KB/s    in 69s

2022-08-22 15:53:10 (416 KB/s) - 'tl_2019_06_tract.zip' saved [29388806]
```

Figure 7-8 shows the files after they are unzipped into the folder hierarchy. Navigate to county-level data, using the same steps you used to locate the state tracts for California.

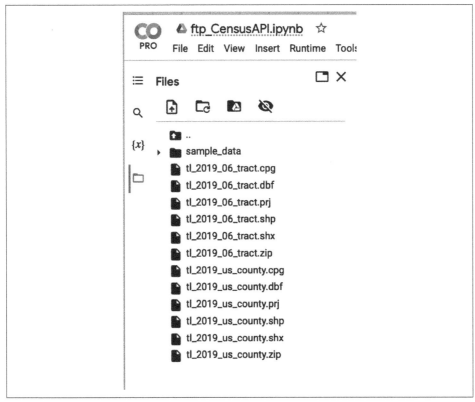

Figure 7-8. Google Colab file structure

Now that the files are unzipped, we can retrieve the FTP files from the web servers:

```
'ftp://ftp2.census.gov/geo/tiger/TIGER2019/COUNTY/tl_2019_us_county.zip'
!wget ftp://ftp2.census.gov/geo/tiger/TIGER2019/COUNTY/tl_2019_us_county.zip
```

Now you can plot the California tracts (`ca_tracts`) using GeoPandas, reading the file from your Colab window. Each file has three dots next to the filename. Right-click on the dots and select Copy Path to copy the link to your files:

```
ca_tracts = gpd.read_file('/content/tl_2019_06_tract.shp')
ca_tracts.plot()
```

The output is shown in Figure 7-9.

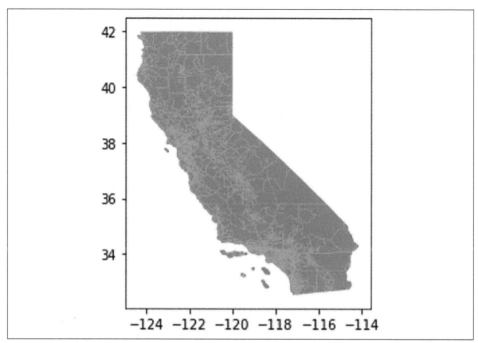

Figure 7-9. California census tracts

Let's take a moment to figure out what the table IDs mean, to help you find the right data.

Accessing Data from the Census API in Your Browser

Next, you'll need to open the US Census Bureau website (*https://oreil.ly/A3Bwq*). In the search window, enter **DP03**, the table ID for census data (Figure 7-10). The ID might seem random, but actually it provides important information. Right away, you can tell from the results that the DP indicates a data profile table. The table IDs (*https://oreil.ly/Fhioi*) describe this table as containing "broad social, economic, housing, and demographic information." The following number in the ID, 03, also refers to selected economic characteristics.

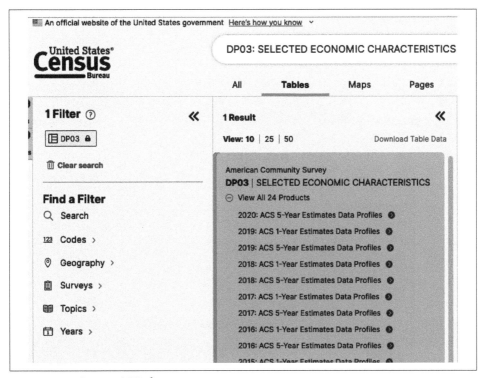

Figure 7-10. Census API: data

Select Enter to see a list of available tables. Figure 7-11 shows the different years and types of ACS surveys available from this search. These are called *data profiles*, and we will touch on them again a little later.

Figure 7-11. Data profile tables available from the US Census Bureau

The Available APIs page on *census.gov* (*https://oreil.ly/tJM7C*) (Figure 7-12) is an important resource that contains publicly available datasets on a wide variety of topics. From here, select American Community Survey (ACS).

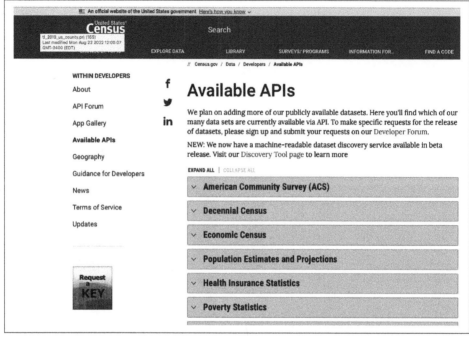

Figure 7-12. Available APIs and other US Census Bureau datasets

Under American Community Survey (ACS), select 1-Year Estimates and scroll down to 2019. Scroll down the list of tables until you see the data profiles. Grab the first link and paste it into your browser. You are going to edit this link and generate a *.csv* file.

 While you're on the Available APIs page, you can request a key for running US Census Bureau service requests. You won't need one for this exercise, but I recommend requesting one and saving it in a reliable location. Look around at the other datasets listed and familiarize yourself with the scope of publicly available data. Another useful US Census Bureau resource is the Developer Forum, an active community that accepts requests for datasets and consultancy.

Using Data Profiles

In the data profile section, you have the example call that you will edit and other details about the data profile tables:

- **Example Call:**

  api.census.gov/data/2019/acs/acs1/profile?get=group(DP02)&for=us:1&key=YOUR_KEY_GOES_HERE
- 2019 ACS Data Profiles Variables [html (*https://oreil.ly/G0PZ0*) | xml (*https://oreil.ly/Iw9CJ*) | json (*https://oreil.ly/ruuB4*)]
- ACS Technical Documentation (*https://oreil.ly/ynqzy*)
- Examples and Supported Geography (*https://oreil.ly/z8LI5*)

Click the link to gather more information. Your first edits will include substituting your table of interest, `DP03`, and removing the placeholder `&key=YOUR_KEY_GOES_HERE` from the URL:

api.census.gov/data/2019/acs/acs1/profile?get=group(DP03)&for=us:1

Then paste the URL into your browser and hit Enter. You will now want to customize the geography portion of your URL to include all metropolitan areas. Select Examples and Supported Geography from the previous page and scroll down to the metropolitan area in the geography hierarchy shown in Figure 7-13.

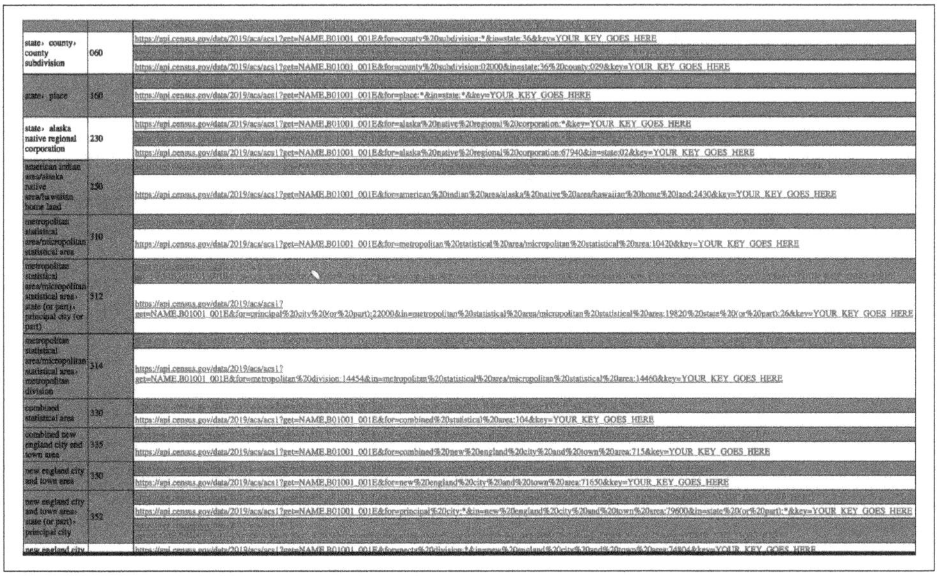

state› county› county subdivision	060	https://api.census.gov/data/2019/acs/acs1?get=NAME,B01001_001E&for=county%20subdivision:*&in=state:36&key=YOUR_KEY_GOES_HERE https://api.census.gov/data/2019/acs/acs1?get=NAME,B01001_001E&for=county%20subdivision:02000&in=state:36%20county:029&key=YOUR_KEY_GOES_HERE
state› place	160	https://api.census.gov/data/2019/acs/acs1?get=NAME,B01001_001E&for=place:*&in=state:*&key=YOUR_KEY_GOES_HERE
state› alaska native regional corporation	230	https://api.census.gov/data/2019/acs/acs1?get=NAME,B01001_001E&for=alaska%20native%20regional%20corporation:*&key=YOUR_KEY_GOES_HERE https://api.census.gov/data/2019/acs/acs1?get=NAME,B01001_001E&for=alaska%20native%20regional%20corporation:67940&in=state:02&key=YOUR_KEY_GOES_HERE
american indian area/alaska native area/hawaiian home land	250	https://api.census.gov/data/2019/acs/acs1?get=NAME,B01001_001E&for=american%20indian%20area/alaska%20native%20area/hawaiian%20home%20land:2430&key=YOUR_KEY_GOES_HERE
metropolitan statistical area/micropolitan statistical area	310	https://api.census.gov/data/2019/acs/acs1?get=NAME,B01001_001E&for=metropolitan%20statistical%20area/micropolitan%20statistical%20area:10420&key=YOUR_KEY_GOES_HERE
metropolitan statistical area/micropolitan statistical area› state (or part)› principal city (or part)	312	https://api.census.gov/data/2019/acs/acs1?get=NAME,B01001_001E&for=principal%20city%20(or%20part):22000&in=metropolitan%20statistical%20area/micropolitan%20statistical%20area:19820%20state%20(or%20part):26&key=YOUR_KEY_GOES_HERE
metropolitan statistical area/micropolitan statistical area› metropolitan division	314	https://api.census.gov/data/2019/acs/acs1?get=NAME,B01001_001E&for=metropolitan%20division:14454&in=metropolitan%20statistical%20area/micropolitan%20statistical%20area:14460&key=YOUR_KEY_GOES_HERE
combined statistical area	330	https://api.census.gov/data/2019/acs/acs1?get=NAME,B01001_001E&for=combined%20statistical%20area:104&key=YOUR_KEY_GOES_HERE
combined new england city and town area	335	https://api.census.gov/data/2019/acs/acs1?get=NAME,B01001_001E&for=combined%20new%20england%20city%20and%20town%20area:715&key=YOUR_KEY_GOES_HERE
new england city and town area	350	https://api.census.gov/data/2019/acs/acs1?get=NAME,B01001_001E&for=new%20england%20city%20and%20town%20area:71650&key=YOUR_KEY_GOES_HERE
new england city and town area› state (or part)› principal city	352	https://api.census.gov/data/2019/acs/acs1?get=NAME,B01001_001E&for=principal%20city:*&in=new%20england%20city%20and%20town%20area:79600&in=state%20(or%20part):*&key=YOUR_KEY_GOES_HERE
new england city		https://api.census.gov/data/2019/acs/acs1?get=NAME,B01001_001E&for=necta%20division:*&in=new%20england%20city%20and%20town%20area:74804&key=YOUR_KEY_GOES_HERE

Figure 7-13. Geography hierarchy

Noticing the * (wildcard option), copy the URL, being sure to delete the `YOUR_KEY_GOES_HERE` section once again. The other option will give you data for a single metropolitan area: the Akron, Ohio, metro area.

Once you are familiar with the URLs and data tables, it takes just a few quick edits to get a usable file with the demographic or economic data of interest. Modify the URL as shown:

https://api.census.gov/data/2019/acs/acs1/profile?get=NAME,DP03_0001E&for=metro-politan%20statistical%20area/micropolitan%20statistical%20area:*

Now paste it into your browser and hit Enter. The file saves in JSON format on my Mac. Instead of saving it as a *.csv* file, which would force you to edit it in Microsoft Excel, run the following lines of code to convert it from *.json* to *.csv*:

```
df = pd.read_json('link to json file')

df.to_csv()
df
```

This will output a formatted *.csv* file.

Creating the Map

To complete our project creating population densities, you will return to the URL for the 2010 census.

Let's begin by looking at the reference URL from the 2010 census:

https://api.census.gov/data/2010/dec/sf1?
get=LSAD_NAME,P001001&for=tract:*&in=state:06

Now that you are familiar with Census API calls, you can break the link down in Python to calculate the densities and customize the tables. The total population in the URL call for the census tracts in California includes table P001001; sf1 refers to the decennial census:

```
# The built out request for the URL  https://api.census.gov/data/2010/dec/
sf1?get=LSAD_NAME,P001001&for=tract:*&in=state:06
HOST = "https://api.census.gov/data"
year = "2010"
dataset = "dec/sf1"
base_url = "/".join([HOST, year, dataset])
predicates = {}
get_vars = ["LSAD_NAME", "P001001"]
predicates["get"] = ",".join(get_vars)
predicates["for"] = "tract:*"
predicates["in"] = "state:06"
r = requests.get(base_url, params=predicates)
```

Now that you've made your request, let's look at the column names for your data:

```
print(r.json()[0])
['LSAD_NAME', 'P001001', 'state', 'county', 'tract']
```

Remember the `index` function. You only need the first row (r) of data to read the headings. You can also create new names:

```
# Create user friendly column names
tract_name = ["tract_name", "tract_pop", "state_fips", "county_fips",
"tract_fips"]
# Reading the json into pandas df
tractdf = pd.DataFrame(columns=tract_name, data=r.json()[1:])
# Changing data types to integer
tractdf["tract_pop"] = tractdf["tract_pop"].astype(int)

tractdf.head()
```

Follow along in the notebook. The following code will select by attributes for all Los Angeles County census tracts, remove any nulls, and create a new `geoid` column. Run the code and verify that the `geoid` column has indeed been added.

First, we need to select attributes from Los Angeles county:

```
onecounty_tractdf = tractdf.loc[tractdf['county_fips'] ==
'037'].copy()Onecounty_tractdf
```

Next, we create the new dataframe that includes our new geoid column:

```
onecounty_tractdf['tract_fips'] = onecounty_tractdf['tract_fips'].str.ljust(6,'0')
onecounty_tractdf['geoid'] = onecounty_tractdf['state_fips'] +
onecounty_tractdf['county_fips'] + onecounty_tractdf['tract_fips']
onecounty_tractdf
```

There are 2,346 records in the dataframe. You can also do a quick count of records with a single line of code:

```
onecounty_tractdf.count()
```

To join the tract-level and county-level files, you will need to select an attribute for the join. To refresh your mind as to the choices, check the column headings with code:

```
ca_tracts.info
```

The join will combine column headings `GEOID` and `geoid` from the two datasets:

```
attr_joined = pd.merge(ca_tracts, onecounty_tractdf, left_on='GEOID',
right_on='geoid')
# Check that all 2345 Census Tracts joined
attr_joined.count()
```

Double-check the coordinate reference system to make sure they match on both the geography file and the census data. ALAND is the area, in square meters. Now you can add the block groups to your map:

```
map = fm.Map(location=[center_y, center_x], zoom_start=10)

# Add Study Area Block Groups to Map
```

```
fm.Choropleth(
    geo_data = ca_prj,
    data=ca_prj,
    columns=['tract_pop','ALAND'],
    key_on= 'feature.properties.tract_pop',
    fill_color='YlGnBu',
    name = 'Population Density',
    legend_name='Population Density'
).add_to(map)
map
```

Population density measures how many people reside within a given area, and it is normally calculated in square kilometers. You can convert the study area to square kilometers right in the code cell:

```
# Create a new column for Census Tract area in square Kilometers

ca_prj['AreaLandKM2'] = (ca_prj['ALAND'] * .000001)

ca_prj[['geoid','TRACTCE','ALAND','AWATER','AreaLandKM2']].head()

ca_prj['ppl_perKM2']=(ca_prj['tract_pop']/ca_prj['AreaLandKM2'])
ca_prj[['geoid','TRACTCE','tract_pop','AreaLandKM2','ppl_perKM2']].head(16)
```

Next, select where you want the center of the map to be:

```
center_x = (ca_prj.bounds.minx.mean() + ca_prj.bounds.maxx.mean())/2
center_y = (ca_prj.bounds.miny.mean() + ca_prj.bounds.maxy.mean())/2
print(f'The center of the data file is located at {center_x} {center_y}')
```

Now you can view the map and the legend (generated automatically) showing population density within Los Angeles County by running the following code:

```
map = fm.Map(location=[center_y, center_x], zoom_start=10)

# Add Study Area Block Groups to Map
fm.Choropleth(
    geo_data = ca_prj,
    data=ca_prj,
    columns=['TRACTCE','ppl_perKM2'],
    key_on= 'feature.properties.TRACTCE',
    fill_color='YlGnBu',
    name = 'Population Density',
    legend_name='Population Density'
).add_to(map)
map
```

The output, your completed map, is shown in Figure 7-14.

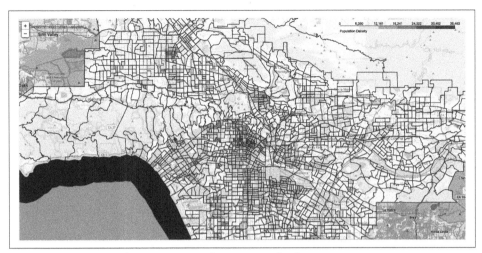

Figure 7-14. Los Angeles County population tracts by density

Importing TIGER files from the Census API FTP site guarantees a single source of location accuracy at a variety of geographies. Here you've looked at census tract levels, but a simple edit to a URL call allows you to change to the level of counties, congressional districts, or municipalities, for example. Although these file types do not include demographic or economic information, you learned how to discover data resources and view them in the notebook. The ability to save JSON files in *.csv* format also means that you can upload them to a GUI, like QGIS, for further exploration.

Summary

GeoPandas is a powerful open source Python project that lets users integrate geographic data with pandas objects and perform spatial operations. It quietly works beneath the surface in geospatial calculations. In this chapter, you have read and written files, selected data, made maps and plots, tackled interactive mapping and geometric manipulations, and explored a few useful packages to include in your future geospatial analyses.

## CHAPTER 8

Data Cleaning

A universal problem when working with data is understanding the completeness of your data. Data engineering depends on the ability to clean, process, and visualize data. Now that you're familiar with the basic functionality of and integration of data with notebook-based code editors, either locally in a Jupyter Notebook or in the cloud with Google Colab, it's time to learn how to clean your data. Data is frequently incomplete (missing), inconsistently formatted, or otherwise inaccurate—problems often called *messy data*. *Data cleaning* is the process of addressing these problems and preparing the data for analysis.

In this chapter, we'll explore some publicly available datasets, finding and cleaning up messes with a few packages that you can load into a Colab notebook. You're going to work with NYPD_Complaint_Data_Historic, a dataset from the open data portal for New York City, NYC Open Data (*https://oreil.ly/W8sNI*), updated on July 7, 2021. I filtered the data for 2020 to make it a little more manageable for viewing and manipulating. You can filter the data based on your data question and export it as a CSV file (*https://oreil.ly/7J4Kj*). This chapter will show you how to manage, remove, update, and consolidate data and process it with a few useful Python packages. Data analysis is only as accurate as the quality of the dataset or database, and this chapter will provide tools to assess and address common inconsistencies.

Checking for Missing Data

If you've ever participated in a data competition, like those available on Kaggle (*https://www.kaggle.com*), you may have noticed that the datasets are engineered to help you focus your attention on a specific task, such as building a visualization. Often, the data has already been cleaned for you. Real life is a little more complicated, and missing data is a persistent problem. Let's see how you can make your data world a little tidier.

With spreadsheets and tabular data formats, evaluating the shape of your data is a straightforward task. A little bit of scrolling can easily reveal rows and/or columns that are missing data. When you review datasets for patterns of missing data, you are actually evaluating them for *nullity*, or the presence of null data (missing values). Geospatial analysis, like analysis in general, often involves multiple tables, so it's important to learn how to identify patterns in the data residing between these tabular datasets.

Uploading to Colab

I am using Google Colab (*https://oreil.ly/J8wam*) for this example, but feel free to use whatever environment you prefer. Upload the NYPD Complaint Data Historic (*https://oreil.ly/kv3Pe*) CSV file to your notebook and click the dots next to the filename. This will give you the path to include in the notebook (Figure 8-1).

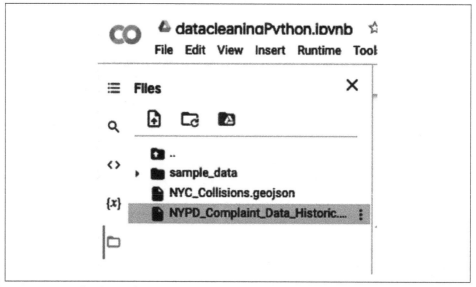

Figure 8-1. Uploading a file to Colab

Missingno (*https://oreil.ly/ps5JI*) is a Python library that detects missing values in a dataset and visualizes how they are distributed within the dataframe, which makes it easier to spot patterns.[1] Missingno has NumPy, pandas, SciPy, matplotlib, and seaborn running under the hood, bringing familiarity to the underlying code snippets. Install missingno:

```
pip install missingno
```

---

[1] Bilogur, Aleksey. 2018. "Missingno: A Missing Data Visualization Suite." *Journal of Open Source Software* 3 (22): 547. *https://doi.org/10.21105/joss.00547*

Depending on the size of your dataset, you can adjust the sample size. Here I've set the sample to 1000. Importing the Python library pandas will allow you to begin a little data cleaning and preparation along the way:

```
import pandas as pd
NYPD = pd.read_csv("/content/NYPD_Complaint_Data_Historic.csv")
import missingno as msno
%matplotlib inline
msno.matrix(NYPD.sample(1000))
```

Once you upload the dataset, you will need to review the data dictionary to see what the column headings indicate.

Converting Spaces to Underscores

You'll often see spaces in the column headings instead of underscores (_). To convert the spaces to underscores, run the following code:

```
df.columns = df.columns.str.replace(' ', '_')
```

In many cases (though not with the current data), you will need to convert some columns in datasets in order to identify them and manipulate them in your code. The df in the code is a placeholder for whatever you named the dataframe (df). In the current example, the df is named NYPD.

The columns listed in our dataframe are shown in Table 8-1. CMPLNT_NUM appears to be the key identifier for each row. This is important because each row is a record of one incident in the dataset.

Table 8-1. Data dictionary for the NYPD Complaint dataset

Field name	Description
CMPLNT_NUM	Randomly generated persistent ID for each complaint
CMPLNT_FR_DT	Exact date of occurrence for the reported event (or starting date of occurrence, if CMPLNT_TO_DT exists)
CMPLNT_FR_TM	Exact time of occurrence for the reported event (or starting time of occurrence, if CMPLNT_TO_TM exists)
CMPLNT_TO_DT	Ending date of occurrence for the reported event, if exact time of occurrence is unknown
CMPLNT_TO_TM	Ending time of occurrence for the reported event, if exact time of occurrence is unknown
ADDR_PCT_CD	Precinct in which the incident occurred
RPT_DT	Date event was reported to police
KY_CD	Three-digit offense classification code
OFNS_DESC	Description of offense corresponding with the key code
PD_CD	Three-digit internal classification code (more granular than the key code)

Field name	Description
PD_DESC	Description of internal classification corresponding with the PD code (more granular than the offense description)
CRM_ATPT_CPTD_CD	Indicator of whether the crime was successfully completed, attempted but failed, or interrupted prematurely
LAW_CAT_CD	Level of offense: felony, misdemeanor, or violation
BORO_NM	Name of the borough in which the incident occurred
LOC_OF_OCCUR_DESC	Specific location of occurrence in or around the premises: inside, opposite of, front of, or rear of
PREM_TYP_DESC	Specific description of the premises: grocery store, residence, street, etc.
JURIS_DESC	Description of the jurisdiction code
JURISDICTION_CODE	Jurisdiction responsible for the incident: either internal, like Police (0), Transit (1), and Housing (2), or external (3), like Correction, Port Authority, etc.
PARKS_NM	Name of New York City park, playground, or greenspace of occurrence, if applicable (state parks are not included)
HADEVELOPT	Name of New York City Housing Authority housing development of occurrence, if applicable
HOUSING_PSA	Development-level code
X_COORD_CD	x-coordinate for New York State Plane Coordinate System, Long Island Zone, NAD 83, units feet (FIPS 3104)
Y_COORD_CD	y-coordinate for New York State Plane Coordinate System, Long Island Zone, NAD 83, units feet (FIPS 3104)
SUSP_AGE_GROUP	Suspect's age group
SUSP_RACE	Suspect's race description
SUSP_SEX	Suspect's sex description
TRANSIT_DISTRICT	Transit district in which the offense occurred
Latitude	Midblock latitude coordinate for Global Coordinate System, WGS 1984, decimal degrees (EPSG 4326)
Longitude	Midblock longitude coordinate for Global Coordinate System, WGS 1984, decimal degrees (EPSG 4326)
Lat_Lon	Geospatial location point (latitude and longitude combined)
PATROL_BORO	The name of the patrol borough in which the incident occurred
STATION_NAME	Transit station name
VIC_AGE_GROUP	Victim's age group
VIC_RACE	Victim's race description
VIC_SEX	Victim's sex description

The variables in a dataset need to be called exactly as they are presented in the dataframe. If they are all capitalized or separated by underscores, you need to be sure to copy them (or you can rename them similarly to how you renamed the columns in the census data in Chapter 7).

Next, there are other considerations to examine in a dataset.

Nulls and Non-Nulls

You'll first need to examine the dataset for missing values. Start by quickly reviewing the data with pandas features such as `NYPD.info()` and `NYPD.describe()` (see Figures 8-2 and 8-4).

Using `.info()` can show you column names, the number of non-nulls, and data types (`Dtype`):

```
NYPD.info()
```

The range index in this dataset is 1,115 entries; therefore, any columns with fewer than 1,115 values are missing data. The column non-null count allows you to view missing data and decide which columns have not captured enough data to provide insights.

This dataset's documentation (*https://oreil.ly/nhqvO*) states that the null values are likely attributable to changes in department forms and to data being collected inconsistently. In addition, if information was unavailable for 2020 (the year I selected to filter the data) or was unknown at the point of collection, it was classified as Unknown/Not Available/Not Reported. It is important to read the supporting documentation to discover the limits of the data, which might in turn limit the questions you can usefully ask.

Data Types

One of your top priorities in data cleaning should be examining the data type of each column in your dataset. The most common data types, which you should definitely be familiar with, are boolean, integer, float, object, and datetime (*https://oreil.ly/WLzh0*).

The most problematic of these data types is the object. Python objects can include a wide variety of data types. Most often they are strings, but you should know that they can also include integers, floats, lists, and dictionaries. In the pandas library, different data types are detected and classified as NaN (the default marker for "not a number") or NaT for missing datetimes. The output of `NYPD.info` includes the data type (`Dtype`), or you can use the `dtypes` DataFrame attribute by appending it with a dot: `NYPD.dtypes`.

Metadata

You can also return information about the shape of the data. This, along with the data type information, is what we call *metadata*, or data about your data. To see the shape of the NYPD data, run:

```
NYPD.shape
(7375993, 35)
```

For simplicity in this example, I filtered the data at the source to include only 2020 data so that we are working with a smaller number of records: 1,115 entries (Figure 8-2).

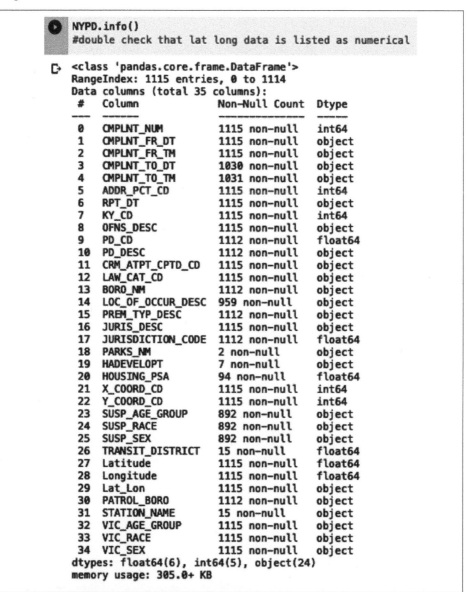

Figure 8-2. Dataframe summary of NYPD.info()

The exact time of each complaint's occurrence, or CMPLNT_FR_TM, is listed as an object or string in Figure 8-2. Pandas will often guess the type of data in a column when a datetime or other data type is assigned as a string or object. This is often complicated by defaults such as the low_memory=True argument. Large files are "chunked," and the file often generates columns with multiple dtypes. I would resist simply switching arguments to False, however. It is much more efficient to simply change the dtype manually, as shown in the following code cell:

```
NYPD = pd.read_csv('/content/drive/MyDrive/NYPD_Complaint_Data_Historic.csv',
                   parse_dates=['CMPLNT_FR_TM'])
NYPD.dtypes.head()

CMPLNT_NUM                int64
CMPLNT_FR_DT             object
CMPLNT_FR_TM     datetime64[ns]

NYPD = NYPD.drop(columns =['CMPLNT_TO_DT','CMPLNT_FR_TM','PARKS_NM','HADEVELOPT',
'HOUSING_PSA'])
```

NYPD.head() will update the columns you retain and confirm that your "dropped" columns no longer appear in the dataframe.

You can eliminate columns with incomplete data if they are not relevant to what you are interested in exploring. However, I strongly discourage you from *deleting* data. Often, the reason for some data being missing is as compelling or significant as the data selected for inclusion.

Let's try this out by identifying a few sample columns to drop from the dataframe.

Summary Statistics

You can examine *summary statistics*, or statistics that summarize your sample data, using .describe().

Look at row counts for another opportunity to identify missing data of the different features in the dataframe. In the output of df.describe(), the NaN values are excluded so that you can see the actual distribution of the data as well as the count of values in each column.

Measures of Central Tendency

A normal distribution is when the mean, mode, and median are all the same value. A *probability* distribution, or *measure of central tendency*, describes the central value in a distribution when the distribution is not normal, as shown in Figure 8-3.

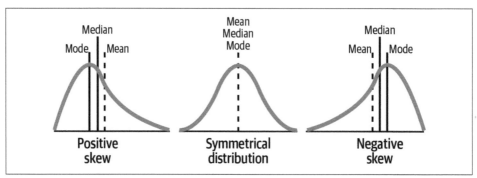

Figure 8-3. A review of measures of central tendency

There are also certain measures that can indicate the presence of skewed distributions and outliers. Figure 8-4 is a snapshot of Energy and Water Data Disclosure for Local Law 84 2021 (Data for Calendar Year 2020) (*https://oreil.ly/vOcv7*), a large dataset that examines energy and water usage in private buildings in New York City. The median or second quartile (Q2), the 50% measure in the table, indicates the *median value*, the point at which half the values are above and half the values are below.

	Property_Id	Largest_Property_Use_Type_-_Gross_Floor_Area_(ft²)	Year_Built	Number_of_Buildings	Occupancy	National_Median_ENERGY_STAR_Score
count	8.410000e+02	8.410000e+02	841.000000	841.000000	841.000000	841.0
mean	6.815681e+06	1.004581e+05	1939.542212	1.084423	96.688466	50.0
std	3.242875e+06	1.579940e+05	33.788712	0.694370	12.066402	0.0
min	7.365000e+03	5.904000e+03	1827.000000	1.000000	0.000000	50.0
25%	4.897515e+06	3.421300e+04	1917.000000	1.000000	100.000000	50.0
50%	6.354826e+06	5.103300e+04	1926.000000	1.000000	100.000000	50.0
75%	6.738968e+06	9.756600e+04	1962.000000	1.000000	100.000000	50.0
max	1.686654e+07	1.976209e+06	2018.000000	14.000000	100.000000	50.0

Figure 8-4. The `df.describe()` function provides descriptive statistics for columns with numerical data

A quick way to look at the shape of your data is to observe where the Q2, or median, is located. If you look at the second column in Figure 8-4, you can determine if the value of the median is closer to 25% or 75%. If it is closer to 25% or Q1, the data is right-skewed, meaning that 25% of the values are below the value for Q1. Think of Q1 as values between the minimum value and the median. The Q3 value is between the median and the maximum values, indicating that 75% of the reported values will be below the Q3. This will result in data with a left skew.

You can have a quick view of the numeric values in your dataset with a df.describe() function. Enter the code snippet into your notebook to view below:

```
df.describe()
```

Plotting a histogram makes it easier to visualize the shape of the data. There is a right-sided tail, or positive skew, as shown in Figure 8-5. This indicates that the mean is greater than the median, which is greater than the mode.

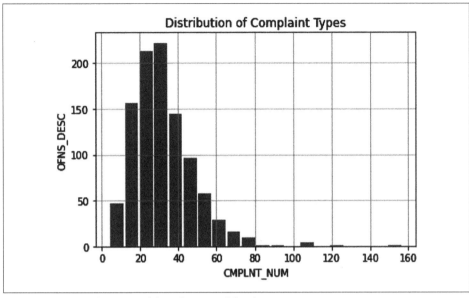

Figure 8-5. Visualization of distribution of the data

Let's return to the NYPD data. We can visualize the data and see its distribution:

```
size, scale = 1000, 10
Complaints = pd.Series(np.random.gamma(scale, size=size) ** 1.5)

Complaints.plot.hist(grid=True, bins=20, rwidth=0.9,
                color='#607c8e')
plt.title('Distribution of Complaint Types')
plt.xlabel('CMPLNT_NUM')
plt.ylabel('OFNS_DESC')
plt.grid(axis='y', alpha=0.75)size, scale = 1000, 10
Complaints = pd.Series(np.random.gamma(scale, size=size) ** 1.5)

Complaints.plot.hist(grid=True, bins=20, rwidth=0.9,
                color='#607c8e')
plt.title('Distribution of Complaint Types')
plt.xlabel('PATROL_BORO')
plt.ylabel('OFNS_DESC')
plt.grid(axis='y', alpha=0.75)
```

The visualization also indicates a rightward skew of the data in the distribution of complaint types (Figure 8-5).

Figure 8-6 is simply a tally of how many values are missing in each column. Knowing when records are missing before analyzing data provides a level of transparency, and it's critical to communicate that information when you share your data visualizations.

```
CMPLNT_NUM                   0
CMPLNT_FR_DT               655
CMPLNT_FR_TM                48
CMPLNT_TO_DT           1704204
CMPLNT_TO_TM           1699541
ADDR_PCT_CD               2166
RPT_DT                       0
KY_CD                        0
OFNS_DESC                18823
PD_CD                     6278
PD_DESC                   6278
CRM_ATPT_CPTD_CD             7
LAW_CAT_CD                   0
BORO_NM                  11329
LOC_OF_OCCUR_DESC      1543800
PREM_TYP_DESC            40745
JURIS_DESC                   0
JURISDICTION_CODE         6278
PARKS_NM               7348330
HADEVELOPT             7029181
HOUSING_PSA            6809283
X_COORD_CD               17339
Y_COORD_CD               17339
SUSP_AGE_GROUP         4795235
SUSP_RACE              3426694
SUSP_SEX               3560008
TRANSIT_DISTRICT       7212494
Latitude                 17339
Longitude                17339
Lat_Lon                  17339
PATROL_BORO               6735
STATION_NAME           7212494
VIC_AGE_GROUP          1638445
VIC_RACE                   309
VIC_SEX                    308
dtype: int64
```

Figure 8-6. NYPD.isna().sum() lists missing values tally for columns in dataset

NYPD.isna() returns the dataframe with Boolean values indicating missing values, as shown in Figure 8-7. In computation, Boolean values have only two possible values: True and False, often represented as 1 and 0. True in this case indicates missing data. At a glance, you can see where the data is missing.

	CMPLNT_NUM	CMPLNT_FR_DT	CMPLNT_FR_TM	CMPLNT_TO_DT	CMPLNT_TO_TM	ADDR_PCT_CD	RPT_DT	KY_CD	OFNS_DESC	PD_CD	PD_DESC	(
0	False	False	False	True	True	False	False	False	False	False	False	
1	False	False	False	False	False	False	False	False	False	False	False	
2	False	False	False	True	True	False	False	False	False	False	False	
3	False	False	False	True	True	False	False	False	False	False	False	
4	False	False	False	True	True	True	False	False	False	True	True	
...	...	...	...	...	...	...	...	...	...	...	...	
7375988	False	False	False	False	False	False	False	False	False	False	False	
7375989	False	False	False	False	False	False	False	False	False	False	False	
7375990	False	False	False	False	False	False	False	False	False	False	False	
7375991	False	False	False	False	False	False	False	False	False	False	False	
7375992	False	False	False	False	False	False	False	False	False	False	False	

7375993 rows × 35 columns

Figure 8-7. Dataframe NYPD.isna()

These are just a few of the options available for exploring a dataframe and understanding the scope of the completeness or nullity of the data.

Now that you've found your missing data, what should you do next?

Replacing Missing Values

An additional reason for examining the dataframe is to review missing values and identify how they're characterized. Once you know how missing values are recorded in the table, you can replace them with a variable of your selection, for example NaN, unknown, or another value of interest, using the na_values parameter:

```
df_test = pd.read_csv("/content/NYPD_Complaint_Data_Historic.csv",
na_values = 'NaN')
print(df)
```

Exploring the dataset often reveals inconsistent reporting of missing values in the data tables. The code snippet will query alternative variables, such as ? and 999.

Visualizing Data with Missingno

In this section, we'll work with missingno again. Missingno can convert tabular data matrices into B*oolean masks*, so called because *masking* returns data based on certain criteria of the underlying manipulated data. The data returns a True/False Boolean according to the indicated criteria. In computation, Boolean values have only two possible values. Missingno marks cells that contain data as True and empty cells as False. It can visualize these in a *nullity matrix* (Figure 8-8), a chart that uses white spaces to visually represent missing data and reveal patterns. Where data is present, it will be shaded.

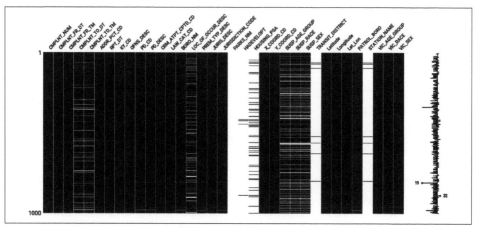

Figure 8-8. Nullity matrix for the NYPD Complaint dataset

In Figure 8-8, each column represents a feature in the data, and each row represents an observation. The sparkline at the far right of the matrix shows how much data is missing in each record.

Do you see any patterns in the missing data here? I notice that the fourth and fifth bars, which represent complaint ending dates and times, have the same pattern: when one of those two values is missing, the other one is too. You can use `msno.bar` to display the same information as a column instead of a simplified matrix:

```
msno.bar(NYPD.sample(1000))
```

The height of the bar equals the nullity, or level of missing data (Figure 8-9). Taller columns indicate that the value is complete (not missing data) or almost complete.

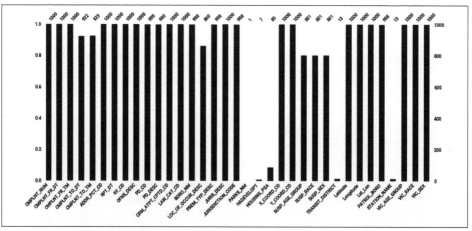

Figure 8-9. Nullity shown by column

Evaluating `msno.heatmap(NYPD)` examines *nullity correlation*, or the absence or presence of a variable, and how strongly this affects the presence of the other variable. Correlation is measured on a scale from –1 to 1, as shown in Figure 8-10. *Zero correlation* means that the presence of one variable appears to be unrelated to whether another variable is present; a correlation score of 1 means that one variable appears whenever the other is present. Negative values indicate *negative correlation:* the presence of one variable indicates that the other variable will not be present.

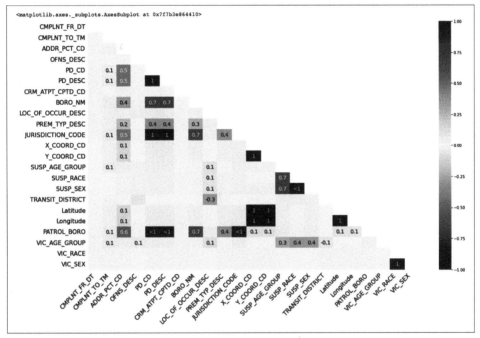

Figure 8-10. Heatmap of nullity data (NYPD)

Values that are either always present or never present are not visualized in the heatmap. For example, there are two columns—RPT_DT (the date of the report to police) and JURIS_DESC (jurisdiction description)—that report no missing data, and they are not visualized in the heatmap. Other values that don't appear to be measuring anything might be generated from erroneous data. A closer examination of the data might reveal something of interest. Pay attention to values that are below 1: the value of –0.3 in Figure 8-10, for instance, is worth a closer look. For example, a value of less than –1 means the correlation is almost 100% negative.

Dendrograms, or tree diagrams, are branching diagrams that visually represent relationships between different groups. The branches are called *clades*, and the terminal ends are called *leaves*. The arrangement of the clades tells us which leaves are most similar to one another based on how close together they are. The height of the branch points indicates how similar or different they are from one another: the greater the height, the greater the difference. The closer to 0 the groupings are in the dendrogram, the more closely the presence of nulls in one column is related to the presence or absence of nulls in the other columns.

The *scipy.cluster.hierarchy.dendrogram* (*https://oreil.ly/3Pxln*) composes clusters by drawing a U-shaped link that joins clades. The dendrogram's orientation defaults to top-down if there are 50 or fewer columns and to left-right when there are more than 50.

In Figure 8-11, the columns denoting the race and sex of the suspect (SUSP_RACE and SUSP_SEX) are more similar to each other than are the columns denoting suspect age group and transit district (SUSP_AGE_GROUP and TRANSIT_DISTRICT).

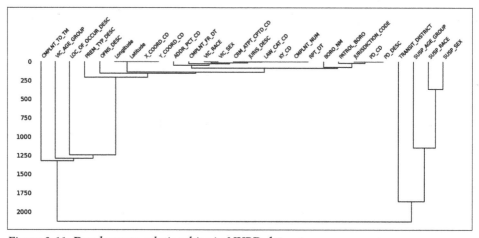

Figure 8-11. Dendrogram relationships in NYPD dataset

To view hierarchical clustering, write the following code snippet into your code cell:

```
msno.dendrogram(NYPD)
```

Dendrograms are useful when focusing on the distribution of data within the dataset. The hierarchical clustering quantitatively estimates the relationship of each data point to every other point in the dataset. Cluster distance is on the vertical axis, and the distinct data points are shown horizontally. The highest level of granularity is at the level of the data sample, and the dendrogram is a way to visualize this quickly.

Mapping Patterns

QGIS is an additional way to view data and look for patterns in the data. The data in Figure 8-12 is filtered to show only burglaries in order to help guide additional insights based on location. Let's look at how to generate such a map.

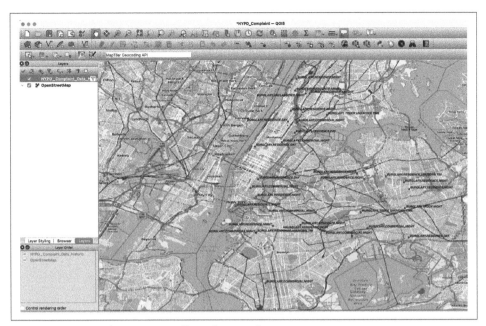

Figure 8-12. Visualizing NYPD filtered crime data on a map

Latitude and Longitude

The dataset has Latitude and Longitude columns. You can convert these to point out features, then use GeoPandas to create a map. Let's use the same file to import data and add a shapefile. First, import pandas:

```
import pandas as pd
```

Next, `read_csv` reads the datafile into your notebook:

```
df = pd.read_csv('/content/drive/MyDrive/NYPD_Complaint_Data_Historic.csv')
```

Upload the file directly to Google Colab if that's what you're using. You can modify the `df` variable to whatever you prefer. The `df.head()` function will return the first rows in each column:

```
df.head()
```

As a best practice, make sure to capture spellings and capitalizations exactly as they appear in your data, and always list longitude before latitude when writing code.

Shapefiles

You will use the Python library Shapely (*https://oreil.ly/KE75Y*) for *computational geometry,* or working with geometric objects like points, curves, and surfaces. When uploading a shapefile, be sure to upload all of the files associated with the shapefile (they'll have the extensions *.dbf, .shx,* and *.prj*) into the same directory. The shapefile in Figure 8-13 is from NYC Open Data Borough Boundaries (*https://oreil.ly/LYjNT*).

Figure 8-13. NYC Open Data borough boundaries

First, import your libraries and packages:

```
!pip install geopandas
import geopandas as gpd
from shapely.geometry import Point, Polygon
import matplotlib.pyplot as plt
street_map = gpd.read_file('/content/tl_2021_36_place.shp')
```

 Python libraries are often abbreviated in code. You've seen that pandas is abbreviated as pd and GeoPandas as gpd. You will notice plt used for matplotlib—don't mistake this for part of a function or argument.

Now Python simply needs to know how to apply the coordinates in a defined space—so you need to define it:

```
# designate coordinate system
crs = {'init': 'epsg:4326'}
# zip x and y coordinates into single feature
geometry = [Point(xy) for xy in zip(df['Longitude'], df['Latitude'])]
# create GeoPandas dataframe
geo_df = gpd.GeoDataFrame(df,
crs = crs,
geometry = geometry)
```

GeoPandas is a *Cartesian coordinate reference system,* which means that each point is defined by a pair of numerical coordinates, such as latitude and longitude in our example. It assigns geographic data to pandas objects. The GeoPandas dataframe has been created from defining our CRS and combining our Longitude and Latitude columns. The `Point` function in Shapely uses the Longitude and Latitude columns to create a new column labeled geometry in Figure 8-14.

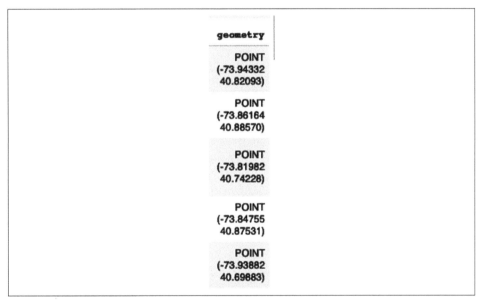

Figure 8-14. Adding a new column, geometry

In the preceding code cell, we defined `geo_df`, and now we can view the newly created geometry column:

```
geo_df.head()
```

Here I have enlarged the `figsize` for better visibility, but it will load more quickly at a smaller size. The default units are inches, but conversions are available (*https://oreil.ly/1PeVm*) for centimeters and pixels as well. The `street_map` is assigned to the axes as that is where we assigned our shapefile:

```
# create figure and define axes, assign to subplot (matplotlib)
fig, ax = plt.subplots(figsize=(30,30))
# add .shp mapfile to axes
street_map.plot(ax=ax, alpha=0.4,color='grey')
# add geodataframe to axes
# assign 'OFNS_DESC' variable to represent coordinates on graph
# add legend
# make data points transparent using alpha
# assign size of points using markersize
geo_df.plot(column='OFNS_DESC',ax=ax,alpha=0.5, legend=True,markersize=10)
# add title to graph
plt.title('Reported Offenses', fontsize=15,fontweight='bold')
# set Latitude and Longitude boundaries for map display
plt.xlim(-74.02,-73.925)
plt.ylim( 40.7,40.8)
# show map
plt.show()
```

Figure 8-15 is a map created from a GeoPandas dataframe. You can filter and edit the reported incidents to create a more tailored map. (Refer to the Jupyter Notebook (*https://oreil.ly/9ADWy*) for additional options.) The `plt.xlim` and `plt.ylim` commands let you select a specific boundary to further edit your projection.

If you would like to select a type of crime from the dataset, use the `df.loc` function to locate all instances. Here is an example showing burglaries:

```
fig, ax = plt.subplots(figsize=(15,15))
street_map.plot(ax=ax, alpha=0.4,color='grey',legend=True,markersize=20)
geo_df.loc[df['OFNS_DESC'] == 'BURGLARY']
geo_df.plot(ax=ax, alpha=0.5,)
plt.xlim(-74.02,-73.925)
plt.ylim( 40.7,40.8)
plt.show()
```

Or perhaps you want to list a few different offenses:

```
df.loc[df['OFNS_DESC'].isin([value1, value2, value3, ...])]
```

Or a combination of parameters:

```
df.loc[(df['OFNS_DESC'] == value) & (df['TRANSIT_DISTRICT'] == value)]
```

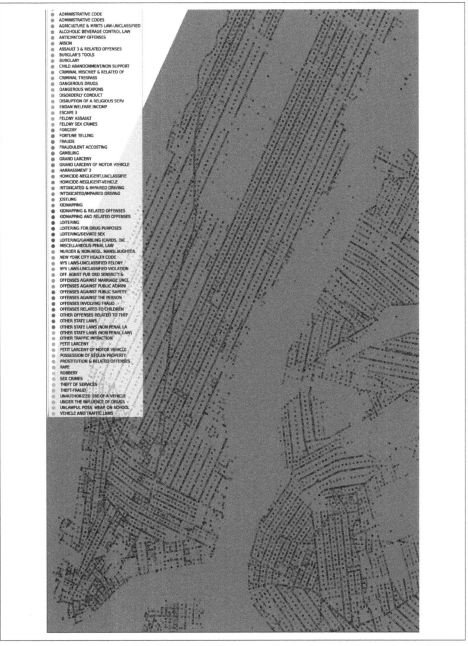

Figure 8-15. GeoPandas dataframe visualized on a map of New York City boroughs

Experiment with asking different questions!

Summary

You have learned a variety of methods useful for cleaning your data. This is important, especially when working with open source data. For example, community-level data is often entered manually, and misspellings, missing dates, and the absence of geometry variables can limit the utility of these valuable resources.

## CHAPTER 9

Exploring the Geospatial Data Abstraction Library (GDAL)

No discussion of open source platforms and libraries would be complete without an introduction to the Geospatial Data Abstraction Library (GDAL) (*https://gdal.org*), a resource for efficiently processing raster and vector geospatial data. Working with both raster and vector data requires a collection of tools, and GDAL operates under the hood of many of the programs you've used in this book, including ArcGIS, QGIS, and Google Earth. When relying on the programmatic ease of a graphical user interface, it is often easy to forget the complexity of combining diverse data types and formats in order to work efficiently with a unified data model. This is why GDAL is an important tool. It simplifies working with geospatial data across a wide range of formats and spatial reference systems.

In this chapter, we'll work with rasters and look at how to use the Spyder IDE to work with GDAL, including how to change your map projection with the `warp` function, work with raster bands, transform files, and create binary masks. We'll also do a quick orientation to three other useful resources for datasets to explore with GDAL: EarthExplorer, Copernicus Open Access Hub, and Google Earth Engine (GEE).

First, I'll show you how to use the command line interface, also known as the *terminal,* to quickly read, transform, and reproject your geospatial data. Why the command line? If you're working with multiple files that require the same functions, you don't have to go through them one by one manually, attempting to recall the necessary steps and processes while working in QGIS or ArcGIS. Using the command line, you can save a simple line of code as a script that you can use to process a wide variety of files and datasets. These scripts also run in QGIS, as you'll see a little later on in the chapter. While working in the command line, I recommend viewing the results in QGIS, or you can refer to the Spyder IDE section once you are comfortable with working in the terminal.

Second, I'll show you how to work with GDAL using an IDE, a tool that consolidates programming and application development. Many beginner Python courses are taught within IDEs, so if you have some background in Python programming, you may be familiar with at least one IDE. There are many IDEs for Python, but in this chapter we'll use one called Spyder. IDEs like Spyder simplify working with and writing code, thanks to features like real-time code introspection (which lets you see directly how the functions and classes are executed and what information they contain), in-line graphics displayed with matplotlib, and my favorite, the *variable explorer*, which provides built-in support for editing and visualizing the objects in your code. IDEs look the same regardless of the operating system you are using, which is handy when your end users are accessing different workflows and resources.

Setting Up GDAL

Historically, the library was referred to as *GDAL/OGR*, with GDAL being the raster side and OpenGIS Simple Features Reference Implementation (OGR) the vector side of the library. The combined libraries are now commonly referred to as *GDAL*. I know there are many data scientists working almost exclusively at the command line, so I want to quickly demonstrate how this is accomplished. For me, working at the command line made it easier to become familiar with the syntax of calling functions in GDAL. Its classes and functions are a little different than those you know from Python, although the syntax will be familiar when you engage with the Python API. GDAL maintains bindings for Python, but this will differ from how you have been working with Python libraries thus far. Even with Python bindings, you will need to know how to call GDAL functions and understand what they do.[1] We'll then get back to Python in the section on working with the IDE.

Installing Spyder

The Spyder IDE is written entirely in Python. It is best known for being used by data scientists, data analysts, and data engineers. Its features include autocompletion and real-time analysis. Let's install it first.

I recommend creating a new environment for the Spyder installation. Installing GDAL into an existing environment is also possible; you can enter **conda env list** to see what environments you already have. You will need to add NumPy and matplotlib as well.

---

1 You may recall that *bindings* are libraries that bridge two programming languages so that a library written for one language can be used in another.

Install Spyder and GDAL in the same environment and directory, substituting your own filepath for mine:

```
spyder-env/Users/bonnymcclain/opt/miniconda3/envs/spyder-env
```

```
conda install -c conda-forge spyder
```

Once you've installed Spyder, you can launch it by writing **spyder** into the terminal.

Installing GDAL

You'll also install GDAL at the command line. Open the terminal and run the following command at the prompt:

```
Conda install -c conda-forge gdal
```

To check if the installation was successful, run:

```
gdalinfo --version
```

You might have to set the path to your installation folder to access it from the terminal. The steps on how you can do this will follow. I also recommend reviewing the GDAL documentation (*https://oreil.ly/ye9fu*).

Working with GDAL at the Command Line

Today, the average computer user works mostly in a GUI: a layer of graphics that makes it easier to visualize files and perform tasks. But the command line has its advantages, including faster processing. There's no need to load a bunch of files, and if you want to repeat an analysis on different datasets, you can save small shell scripts to perform repeated actions.

Understanding Your Directories

If you are new to working in the terminal, spend a few minutes understanding your directories. You need to know your current directory and how to move between different directories, depending on where your dataset is located. The command pwd, for "print working directory," will show you the filepath of your current location, such as */Users/bonnymcclain*.

In this example, I am using the minimal data science environment (minimal_ds), as referenced in this article (*https://oreil.ly/BJUAC*), which advocates for installing the leaner version of Anaconda, Miniconda, including steps for setting up your environment:

```
(minimal_ds) MacBook-Pro-8:~ bonnymcclain$ pwd
/Users/bonnymcclain
```

If you get lost:

cd *(change directory)*
  Tells you your current directory

cd..
  Moves up one directory level

pwd *(print working directory)*
  Shows you the full pathname of the directory you are in

ls *(list)*
  Lists the folders in your working directory

Activate your preferred environment, conda activate minimal_ds, or the Spyder environment you just created for working with GDAL. I would like to work with files in my TIFFs folder so I have entered cd to make that my current directory. You can now see TIFFs added as my working directory:

```
(minimal_ds) MacBook-Pro-8:~ bonnymcclain$ cd TIFFs
(minimal_ds) MacBook-Pro-8:TIFFs bonnymcclain$ ls
```

To move between different directories, use ls/ to dig deeper into your directory structure. For example, if you want to work with your Downloads folder (assuming it isn't your current directory), use the following code to explore a folder within your Downloads directory:

```
ls/Downloads/another folder

(minimal_ds) MacBook-Pro-8:~ bonnymcclain$ cd TIFFs
(minimal_ds) MacBook-Pro-8:TIFFs bonnymcclain$ ls
```

Spend a little time moving in and out of directories, as understanding the structure is the most important lesson for working in the terminal with the GDAL package. Again, you will need to understand the structure when accessing packages, libraries, and files in the Spyder IDE.

If you are in the correct directory, you will see the directory name after the colon and right before your username. This will output a list of the files in this directory:

```
TIFFs bonnymcclain$ ls
```

I have included a section on resources for creating your own files using GEE (*https://oreil.ly/6dRtp*) for following along. Copernicus Open Access Hub (*https://oreil.ly/Z8FSL*) and EarthExplorer (*https://oreil.ly/WvlrA*) are also sources of raster data but are often rendered slowly due to the sizes of larger files.

Now that your installation is working, you are ready to learn a few commands to help you get started. Here is the GDAL documentation (*https://oreil.ly/ye9fu*) to help you through any lingering problems with the downloading and installation.

Editing Your Data with GDAL

The `gdalinfo` command line parameters demonstrate the information available to you about your raster dataset. You can see the parameters by using the command `gdalinfo` and supplying the name of a file within the active directory. Enter this code in the terminal:

```
gdalinfo [--help-general] [-json] [-mm] [-stats | -approx_stats] [-hist] [-nogcp]
[-nomd]
        [-norat] [-noct] [-nofl] [-checksum] [-proj4]
        [-listmdd] [-mdd domain|`all`]* [-wkt_format WKT1|WKT2|...]
        [-sd subdataset] [-oo NAME=VALUE]* [-if format]* datasetname
```

In this example I've selected a saved file from a GEE data snippet. You can learn more about the command line parameters in the documentation (*https://oreil.ly/lu9Z0*).

Enter the following code, substituting the name of your *.tif* file in your directory (notice that parentheses are not needed):

```
(minimal_ds) MacBook-Pro-8:TIFFs bonnymcclain$ gdalinfo Sentinel2_RGB20200501.tif
```

Here is a snippet of the output:

```
(minimal_ds) MacBook-Pro-8:TIFFs bonnymcclain$ gdalinfo Sentinel2_RGB20200501.tif
Driver: GTiff/GeoTIFF  This is the format of the saved file
Files: Sentinel2_RGB20200501.tif
Size is 5579, 4151
Coordinate System is:
PROJCRS["WGS 84 / UTM zone 29N",
    BASEGEOGCRS["WGS 84",
        …
Pixel Size = (10.000000000000000,-10.000000000000000)
Metadata:
  AREA_OR_POINT=Area
Image Structure Metadata:
  COMPRESSION=LZW
  INTERLEAVE=PIXEL
Corner Coordinates:
Upper Left  (566320.000, 4133200.000) (8d15' 4.53"W, 37d20'35.24"N)
Lower Left  (566320.000, 4091690.000) (8d15'17.78"W, 36d58' 8.31"N)
Upper Right (622110.000, 4133200.000) (7d37'17.50"W, 37d20'14.86"N)
Lower Right (622110.000, 4091690.000) (7d37'41.89"W, 36d57'48.20"N)
Center      (594215.000, 4112445.000) (7d56'20.39"W, 37d 9'13.16"N)
```

You can see from the second line that the format of the saved file is GTiff/GeoTIFF.

To see a list of other formats, enter `gdal_translate --formats` in the command line. You can also see the size of the file, coordinate system, pixel size, and coordinates.

Here is a section of the output that shows information about the color bands:

```
Band 1 Block=256x256 Type=Float32, ColorInterp=Gray
  Description = B2
Band 2 Block=256x256 Type=Float32, ColorInterp=Undefined
  Description = B3
Band 3 Block=256x256 Type=Float32, ColorInterp=Undefined
  Description = B4
```

Notice the color interpretation of the bands in the Sentinel satellite data. They are set as `Gray` for B2 and `ColorInterp=Undefined` for the remaining two bands. Because the metadata was not downloaded, you'll need to help GDAL interpret these bands.

Since this is Sentinel 2 data (*https://oreil.ly/mO6iH*), you know that the bands are blue (B2), green (B3), and red (B4). You're going to edit the dataset right in place, using the command `gdal_edit.py`, followed by the options (such as `colorinterp`), and finally the value (here, the color value of the band). The last step is to provide the input file, *Sentinel2_RGB20200501.tif*:

```
gdal_edit.py -colorinterp_1 blue -colorinterp_2 green -colorinterp_3 red
Sentinel2_RGB20200501.tif
```

If your function requires an output file, you will also add *output.tif*. Only the input file is needed here.

When you rerun `gdalinfo Sentinel2_RGB20200501.tif`, you will see that the color bands have been updated and now show as blue, green, and red.

The Warp Function

You can also change the raster projection by learning to use the `gdal_warp` function. (Once you're familiar with the `warp` function, you can use it to learn other common functions; to explore them, I recommend starting with the GDAL documentation for the Python API (*https://oreil.ly/AKsHX*).)

In the following code, the argument `-t_srs` specifies the coordinate system you are targeting. Every geographic coordinate system, for example, is assigned a unique EPSG code. Where the *xxxxx* appears, enter the EPSG code you would like to change to. The input is the raster file, and the output is the renamed modified file, shown here as *output_rd.tif*:

```
gdalwarp -t_srs EPSG: xxxxx  Sentinel2_RGB20200501.tif output_rd.tif
gdalinfo Sentinel2_RGB20200501.tif
```

Your output will demonstrate the new projection.

Capturing Input Raster Bands

Next, you'll use the command line with GDAL to capture image statistics for the projected bands:

```
(minimal_ds) MacBook-Pro-8:TIFFs bonnymcclain$ gdalinfo -stats srtm_41_19_4326.tif
```

An excerpt from the output:

```
STATISTICS_MAXIMUM=640
STATISTICS_MEAN=256.70790240722
STATISTICS_MINIMUM=27
STATISTICS_STDDEV=119.8746675927
STATISTICS_VALID_PERCENT=100
```

Why would you capture input raster bands for statistical analyses? Mostly for classification: to identify clusters when looking at different locations. For example, band characteristics help define whether you are observing vegetation, a body of water, or perhaps a residential area.

I mentioned earlier in the chapter that GDAL runs under the hood in QGIS as well as in a wide variety of other programs. Figure 9-1 demonstrates the code running within the QGIS platform, uploading your *.tif* file(s), and selecting raster information.

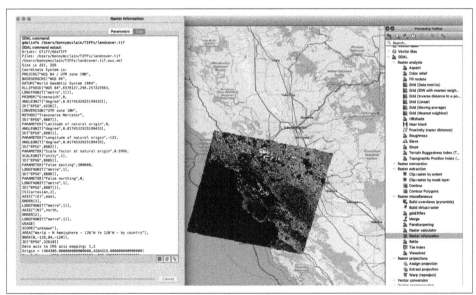

Figure 9-1. gdalinfo displayed in QGIS

Now that you have a taste of working with GDAL in the command line, I encourage you to explore more of the tasks available. These skills are useful for working with and editing some of the core algorithms of geospatial data and applications. The ability to transform data across platforms and diverse formats is important, especially when you're working with a wide variety of vector and raster data formats.

Next, we will work with GDAL integrated with Python.

Working with the GDAL Library in Python

When you launch Spyder (again, by typing **spyder** into the terminal), the Spyder console will open in a new browser window, as seen in Figure 9-2, except it will be blank.

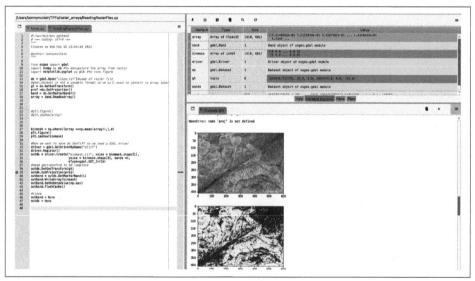

Figure 9-2. The Spyder console

Getting Oriented in Spyder

On the left side of Figure 9-2 is Spyder's script editor where you create and open files. (You can arrange the panels and consoles however you like.) The two consoles on the right are where you can explore the variables you generate and locate files. The panel that shows the images is the plots pane. You can write simple code in the bottom console. For help, select the Help tab in the upper window.

You can run code in the Python console or from the Editor. You can also use the Python console to control Spyder's debugger. The images and figures you generate will be visible in the plots plane or embedded in the console.

The variable explorer, shown in the top right pane in Figure 9-2, is one of my favorite reasons for working in an IDE. These are the objects generated when you run your code. Click on a variable to explore it in more detail.

Each console window also has a "hamburger" icon, or ☰, which expands into a menu (Figure 9-3) where you can find additional information about the pane.

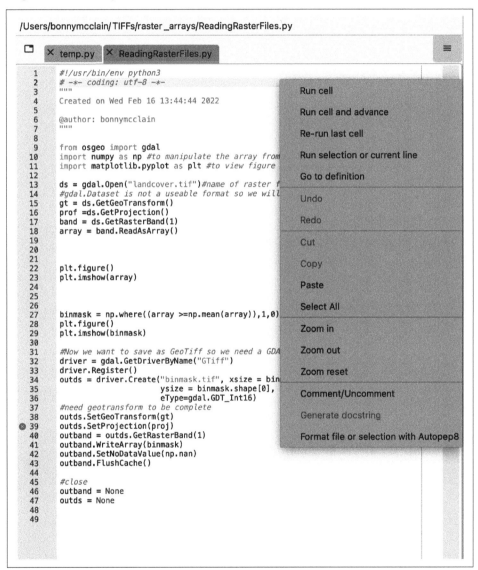

Figure 9-3. Working in console directories

These options are customizable. I suggest heading over to the Spyder website (*https://oreil.ly/SGOgy*) for details about the interface as well as to explore all of the available panes. If you would like to try Spyder before installing, there is also a Binder option (*https://mybinder.org*) that runs in your browser.

Although it's not necessary, I like to start a new project when working in Spyder, as demonstrated in Figure 9-4. This simplifies returning to your prior files, regardless of your current working directory. The project pane will also become visible. It reminds me of the ease of QGIS, since the files are easily within reach.

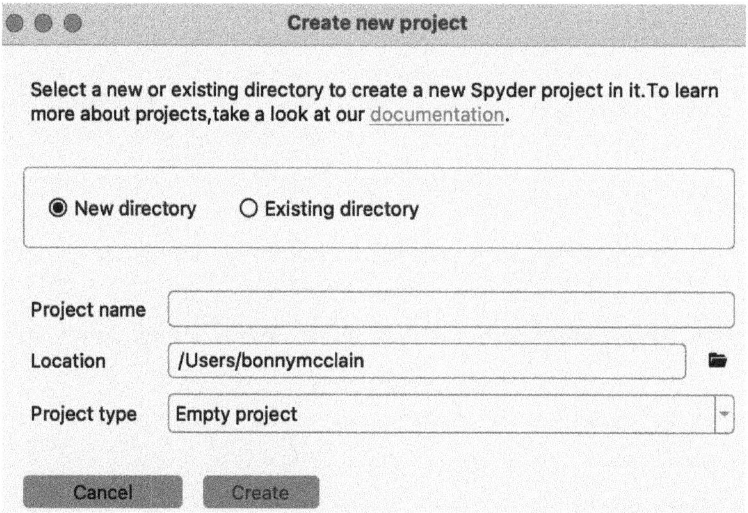

Figure 9-4. Creating a new project in Spyder

The Project pane can also integrate with Git for version control. Explore the Spyder file documentation to learn (*https://oreil.ly/8M3ZI*) how to set up the functionality with your repository.

Exploring Your Data in Spyder

You're just a few lines of code away from exploring your raster data within an IDE. The first step is to import the necessary packages: GDAL, NumPY, and matplotlib:

```
from osgeo import gdal
import numpy as np #to manipulate the array from raster
import matplotlib.pyplot as plt #to view figure
```

NumPY will manipulate the array you are creating from the raster file, and matplotlib will allow you to view the file. The raster programs you can explore are included in the GDAL documentation (*https://oreil.ly/rxgLu*).

Utility scripts belong to a *utility class*, which is a collection of related methods available across the application. The Python utility scripts are located inside the `osgeo_utils.samples` submodule shown in the following code:

```
ds = gdal.Open("slope.tif")
gt = ds.GetGeoTransform()
proj =ds.GetProjection()
band = ds.GetRasterBand(1)
array = band.ReadAsArray()
```

If you are not sure where to find *.tif* files, skip ahead to "Exploring Open Source Raster Files" on page 200, where I will review how to find raster files to explore. These are simple digital elevation models (DEMs), where each point or pixel has an elevation value. Typically, these are represented as DEM *.tif* files. If you've been exploring as you work through this book, your Downloads folder probably harbors quite a few by now.

The code scripts you will be writing into the editor are explained in the text; additional information is available from the GDAL Python API (*https://oreil.ly/lHQri*).

You can call your variable anything, but for simplicity I am using ds to represent the dataset. After importing your *.tif* file and creating the variable ds, you will see the variable populate in the variable explorer. The format of the ds file is `gdal.Dataset`.

Transforming Files in GDAL

You will need to transform your *.tif* file from the table format that you see in Figure 9-6 to geographic coordinates, so next you'll define gt, the geotransform.

The six coefficients in Figure 9-5, read from top to bottom, correspond to:

0

The *x*-coordinate of the upper-left corner of the upper-left pixel (262846.525725)

1

Pixel resolution, west to east (25.0)

2

Row rotation (usually 0)

3

The *y*-coordinate of the upper-left corner of the upper-left pixel (4464275.0)

4

Column rotation (again, usually 0)

5

The northwest pixel resolution and height (usually negative for a north-up image), -25.0

Inde▲	Type	Size	Value
0	float	1	262846.525725
1	float	1	25.0
2	float	1	0.0
3	float	1	4464275.0
4	float	1	0.0
5	float	1	−25.0

Figure 9-5. Geotransforms to georeferenced coordinates

The projection information is visible in the variable explorer (Figure 9-2) but is shown larger in Figure 9-6. In the Universal Transverse Mercator (UTM), the projection is UTM zone 30 N; the EPSG code is EPSG:32630.

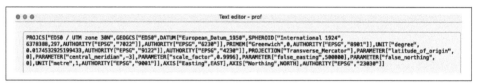

```
PROJCS["ED50 / UTM zone 30N",GEOGCS["ED50",DATUM["European_Datum_1950",SPHEROID["International 1924",
6378388,297,AUTHORITY["EPSG","7022"]],AUTHORITY["EPSG","6230"]],PRIMEM["Greenwich",0,AUTHORITY["EPSG","8901"]],UNIT["degree",
0.0174532925199433,AUTHORITY["EPSG","9122"]],AUTHORITY["EPSG","4230"]],PROJECTION["Transverse_Mercator"],PARAMETER["latitude_of_origin",
0],PARAMETER["central_meridian",-3],PARAMETER["scale_factor",0.9996],PARAMETER["false_easting",500000],PARAMETER["false_northing",
0],UNIT["metre",1,AUTHORITY["EPSG","9001"]],AXIS["Easting",EAST],AXIS["Northing",NORTH],AUTHORITY["EPSG","23030"]]
```

Figure 9-6. The projection of the bands, as read by GDAL

GetRasterBand fetches the bands into the dataset. To determine how many bands you have, write into the console:

```
ds.RasterCount
```

This outputs: 1.

Enter the number of bands into the function:

```
band = ds.GetRasterBand(1)

array = band.ReadAsArray()
```

Exploring the array variable, you can see that Python has read the GDAL file into a NumPY array. The array shares elevations from our DEM, as shown in Figure 9-7.

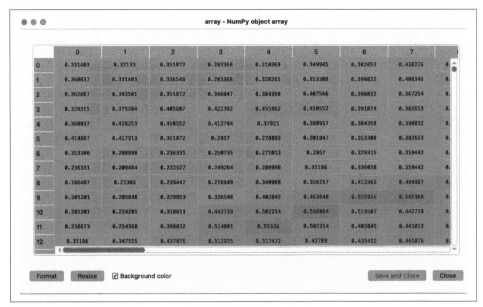

Figure 9-7. NumPY object array of DEM .tif file

Using the Binmask in GDAL

Masking is useful for clipping boundaries or values to a specific range. A *binary mask* (often called *binmask)* is a tool that filters out elevation values above or below a certain number (for instance, if you want to look only at locations that are below sea level). If you want to save everything that is equal to or larger than the mean, you will assign it a value of 1; otherwise, it will be a 0. The output is shown in Figure 9-8. The dark values are at lower elevation.

The `binmask` variable in the code refers to the NumPy (`np`) function that returns the elements in the array based on a condition, such as the `mean`:

```
binmask = np.where((array >=np.mean(array)),1,0)
plt.figure()
plt.imshow(binmask)
```

To save this data as a GeoTIFF file, you will need a GDAL driver to support your chosen raster file format. You can view the long list of different drivers available, including for GeoTIFF, in the GDAL documentation (*https://oreil.ly/MCFi9*). To get the shape of your binmask, you only need to register the driver and run the following code:

```
binmask.shape
```

This outputs (410, 601).

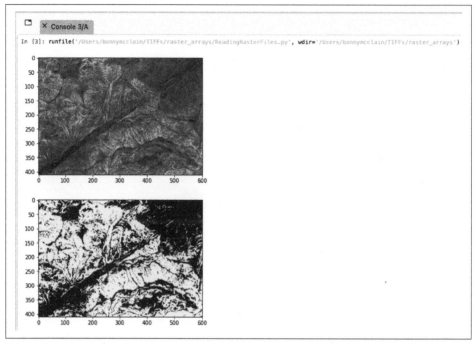

Figure 9-8. Binary mask output

Recall how to use the index function in Python. The binmask shape is 410 rows and 601 columns. In the following code snippet, xsize refers to the number of columns. You call columns by index, in this case as [1], to get a count (in this case, 601). You can do the same with ysize and the index [0] to grab the number of rows (in this case, 410):

```
#Now we want to save as GeoTiff so we need a GDAL driver
driver = gdal.GetDriverByName("GTiff")
driver.Register()
outds = driver.Create("binmask.tif", xsize = binmask.shape[1],
                      ysize = binmask.shape[0], bands =1,
                      eType=gdal.GDT_Int16)
```

gdalconst defines the data type in the image. Since you are defining the shape of the binmask with 0 and 1, you need to set the value as an integer, which you'll do using gdal.GDT_Int16. Now you only need to print the datatype property.

The last step is another geotransform. The settings can stay the same since you didn't change anything. Once you close the files, they will be available for you to use. If you forget, you will not be able to use them. You also don't need to read the array, since you haven't stored it anywhere yet. `WriteArray(binmask)` will provide the output:

```
#need geotransform to be complete
outds.SetGeoTransform(gt)
outds.SetProjection(proj)
outband = outds.GetRasterBand(1)
outband.WriteArray(binmask)
outband.SetNoDataValue(np.nan)
outband.FlushCache()

#close
outband = None
outds = None
```

To recap, you've identified and generated an image from a DEM and converted the output to a GeoTIFF file. The raster image is saved along with any information or metadata about the image's location on the Earth's surface at the pixel level.

The Complete Script

Here is the code in its entirety; uncomment the print option when you want to view the figure:

```
from osgeo import gdal
import numpy as np
import matplotlib.pyplot as plt

ds = gdal.Open("slope.tif")
gt = ds.GetGeoTransform()
proj=ds.GetProjection()
band = ds.GetRasterBand(1)
array = band.ReadAsArray()

#plt.figure()
#plt.imshow(array)

binmask = np.where((array >=np.mean(array)),1,0)
plt.figure()
plt.imshow(binmask)


driver = gdal.GetDriverByName("GTiff")
driver.Register()
outds = driver.Create("binmask.tif", xsize = binmask.shape[1],
                      ysize = binmask.shape[0], bands =1,
                      eType=gdal.GDT_Int16)
```

```
outds.SetGeoTransform(gt)
outds.SetProjection(proj)
outband = outds.GetRasterBand(1)
outband.WriteArray(binmask)
outband.SetNoDataValue(np.nan)
outband.FlushCache()

#DON'T FORGET TO CLOSE FILE

outband = None
outds = None
```

Exploring Open Source Raster Files

The open source geospatial community has a huge arsenal of publicly available datasets (*https://oreil.ly/GIbyH*) to continue your learning and fuel current and future projects. This book is intended to be a living resource that invites additional learning and skill development. So in the next few sections, I'll provide the basics you need to get started with exploring some of these resources, rather than full exercises. The goal is to get you to a point where you can start exploring for yourself.

USGS EarthExplorer

The USGS hosts one of the largest free repositories of satellite and aerial imagery data, called EarthExplorer.

You will need to register for a free account at EarthExplorer (*https://oreil.ly/WvlrA*). If you have a compressed ZIP file that includes a shapefile (*.shp*), you can upload it using the KML/Shapefile Upload button in the upper left corner. Free open source data resources often offer the option to download a zipped shapefile. All of the accompanying files must be uploaded along with the shapefile.

To work with EarthExplorer, you can upload a shapefile you want to work with. Figure 9-9 shows a file I uploaded called "Bodega Marine Laboratory and Reserve." When I draw a polygon (shown in red) around the area I want to look at, I can download the GeoTiff file. You can change the coordinates to reshape the polygon.

Figure 9-9. Uploading a shapefile to EarthExplorer

In addition to uploading a shapefile, there are a few other ways to create an image:

- Zoom to an area you wish to explore and draw a polygon or circle
- Search for an address (Figure 9-10)
- Double-click the map and select the Use Map button
- Select a date range

You can also enter a date range or a cloud cover range in the menu shown in Figure 9-10. For this exercise, use the method of your choice to follow me to the Bodega Marine Laboratory and Reserve. Enter the location into the search criteria.

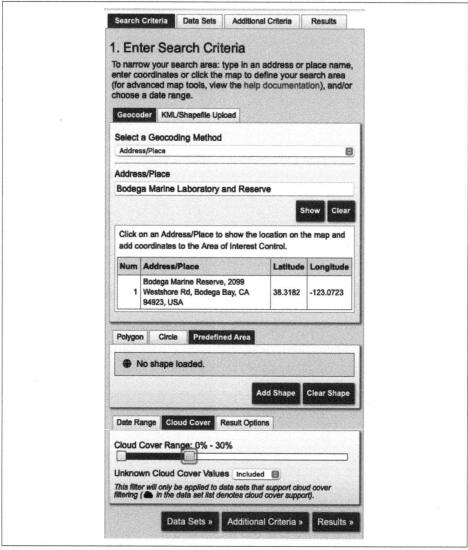

Figure 9-10. Searching for satellite images in your region of interest

Once you're at the Bodega Marine Laboratory and Reserve, select Digital Elevation Data and then SRTM Void Filled, as shown in Figure 9-11. *SRTM* stands for Shuttle Radar Topography Mission. Void-filled SRTMs have additional processing to fill in missing data.

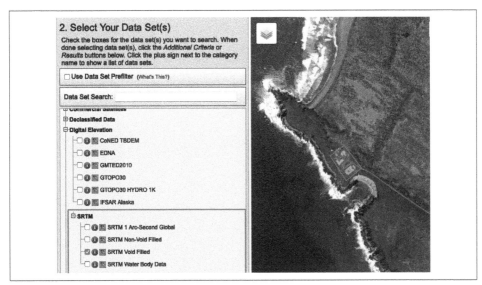

Figure 9-11. Digital elevation SRTM data in EarthExplorer

If there are any images that meet your criteria, they will load as search results (Figure 9-12). Once you adjust your parameters, you will see thumbnails of the datasets. Find the best dataset for your needs and download the GeoTIFF. Save it to the folder within your working directory or use the absolute (complete) filepath to upload it into your Spyder console.[2]

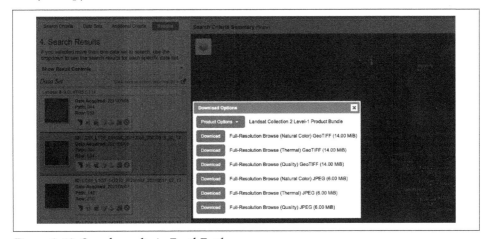

Figure 9-12. Search results in EarthExplorer

---

2 The *relative filepath* is the path relative to your current working directory; the *absolute filepath* is the path provided from your root directory.

Copernicus Open Access Hub

The next data resource I want to show you is Copernicus Open Access Hub (*https://oreil.ly/WR178*). You'll access it similarly to how you accessed EarthExplorer, navigating the settings within a dashboard. While Copernicus's interface is perhaps less intuitive than EarthExplorer's, it offers some great data. Try searching for Sentinel satellite data (Figure 9-13).

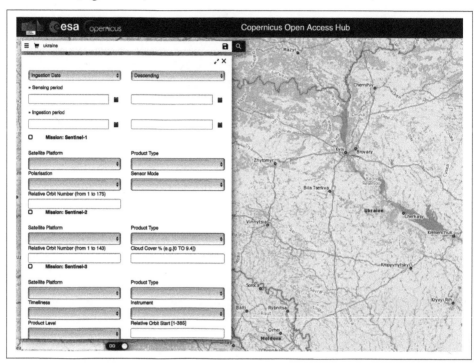

Figure 9-13. Copernicus Open Access Hub Sentinel data

Google Earth Engine

You learned about GEE in Chapter 8, so here I'll just briefly touch on how to use it with GDAL. GEE data is available in the cloud on a planetary scale, so you don't have to download it. You can even use functions such as clipping satellite data.

Run search for GEE DEM files. Find Earth Engine snippet *ee.Image("USGS/sDEP/10m")*, shown in Figure 9-14. You can copy the JavaScript code and paste it into the GEE console.

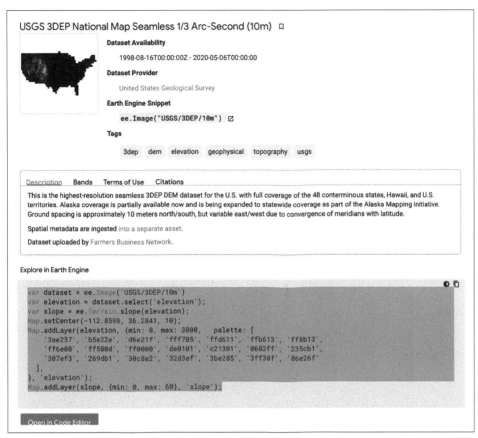

```
USGS 3DEP National Map Seamless 1/3 Arc-Second (10m)  ⌐

      Dataset Availability

          1998-08-16T00:00:00Z - 2020-05-06T00:00:00

      Dataset Provider

          United States Geological Survey

      Earth Engine Snippet

          ee.Image("USGS/3DEP/10m")  ☑

      Tags

          3dep   dem   elevation   geophysical   topography   usgs


Description   Bands   Terms of Use   Citations

This is the highest-resolution seamless 3DEP DEM dataset for the U.S. with full coverage of the 48 conterminous states, Hawaii, and U.S.
territories. Alaska coverage is partially available now and is being expanded to statewide coverage as part of the Alaska Mapping Initiative.
Ground spacing is approximately 10 meters north/south, but variable east/west due to convergence of meridians with latitude.

Spatial metadata are ingested into a separate asset.

Dataset uploaded by Farmers Business Network.


Explore in Earth Engine

var dataset = ee.Image('USGS/3DEP/10m')
var elevation = dataset.select('elevation');
var slope = ee.Terrain.slope(elevation);
Map.setCenter(-112.8598, 36.2841, 10);
Map.addLayer(elevation, {min: 0, max: 3000,   palette: [
    '3ae237', 'b5e22e', 'd6e21f', 'fff705', 'ffd611', 'ffb613', 'ff8b13',
    'ff6e08', 'ff500d', 'ff0000', 'de0101', 'c21301', '0602ff', '235cb1',
    '307ef3', '269db1', '30c8e2', '32d3ef', '3be285', '3ff38f', '86e26f'
  ],
}, 'elevation');
Map.addLayer(slope, {min: 0, max: 60}, 'slope');

Open in Code Editor
```

Figure 9-14. GEE catalog of DEM files

In the GEE console, select Run. This will generate the map shown in Figure 9-15. You can select a polygon and create a geometry import. The layers panel will allow you to change the opacity of a layer or toggle layer displays on and off. Simply save the file as a GeoTIFF, and you have another option for a DEM file.

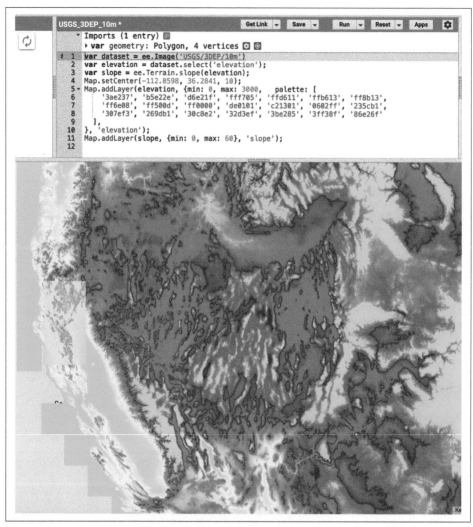

```
USGS_3DEP_10m *                          Get Link  ▾    Save  ▾      Run  ▾    Reset  ▾    Apps   ⚙
  ↻          ▾ Imports (1 entry) ▤
             ▸ var geometry: Polygon, 4 vertices ⊡ ◉
  i  1   var dataset = ee.Image('USGS/3DEP/10m')
     2   var elevation = dataset.select('elevation');
     3   var slope = ee.Terrain.slope(elevation);
     4   Map.setCenter(-112.8598, 36.2841, 10);
     5 ▾ Map.addLayer(elevation, {min: 0, max: 3000,    palette: [
     6        '3ae237', 'b5e22e', 'd6e21f', 'fff705', 'ffd611', 'ffb613', 'ff8b13',
     7        'ff6e08', 'ff500d', 'ff0000', 'de0101', 'c21301', '0602ff', '235cb1',
     8        '307ef3', '269db1', '30c8e2', '32d3ef', '3be285', '3ff38f', '86e26f'
     9     ],
    10   }, 'elevation');
    11   Map.addLayer(slope, {min: 0, max: 60}, 'slope');
    12
```

Figure 9-15. Google Earth Engine DEM .tif file

GDAL can be complex, but it's worth learning how to use this resource library to extend your geospatial skills.

Summary

You have observed how to work in both a terminal and an IDE, both useful in their own respects specific to personal preference and often accessibility. These skills highlight the utility of both options as well as a powerful interface relied on by cross-platform applications in both the enterprise (Esri) and open source communities.

Remember, when working with geospatial data and a broad array of available tools, especially in the open source community, it is important to read the user documentation to grow your skills and your ability to interact with your data. Don't hesitate to reach out to the community with questions or insights.

## CHAPTER 10

Using Python to Measure Climate Data

Developing technical skills and pathways for learning Python and geospatial analysis is important, but unless you provide context or create a narrative to share, it's all simply data on a shelf.

In this final chapter, you will explore three approaches to exploring time-series data by accessing satellite image layers from Landsat (*https://oreil.ly/9bHFj*), China–Brazil Earth Resources Satellite (CBERS) (*https://oreil.ly/PM7Lt*), and Sentinel (*https://oreil.ly/1Kym6*). You will use your geospatial analysis skills to examine questions about climate change and deforestation.

Spatial modeling is a crucial tool for forecasting, predicting, and monitoring the real-time status of global temperature increases and deforestation, which in turn helps us anticipate the consequences of these phenomena and potentially intervene or prepare for them.

Three examples are presented to highlight some powerful Python packages: Xarray, Web Time Series Service (WTSS), and Forest at Risk (FAR). Although these may appear to be new tools, you have been introduced to many of their dependencies in earlier chapters. The last example is a deeper dive into the statistical power of packages designed for predictive modeling, which you'll use in analyzing deforestation. You can run code in the accompanying notebook (*https://oreil.ly/9ADWy*), since complete explanations of everything in it is beyond the scope of this book.

Example 1: Examining Climate Prediction with Precipitation Data

Spatial analysis often relies on multidimensional data analysis. Think of a gridded dataset as resembling a cube. In Python (and in computer programming in general), arrays store lists of data. The objects in the list can be referenced individually or collectively. This is important when calling a Python array because you can access each item by its index number.

Multidimensional and N-dimensional arrays, or *tensors*, are displayed in NumPy ndarrays. Think of a tensor as a container of data or information. In Python, NumPy (*http://www.numpy.org*) provides the fundamental data structure and API for working with raw ndarrays.

You will be working with real-world datasets that encode information about how the array's values map to locations. The data you are working with is labeled with encoded information such as timestamps, coordinates, elevation, land cover, and precipitation.

Goals

The mission of the US government's National Oceanic and Atmospheric Administration (NOAA) (*https://oreil.ly/NKI29*) is to "predict changes in climate, weather, oceans, and coasts" and to inform and address urgent societal and environmental impacts from extreme weather events. In this exercise, you will work with a publicly available dataset to analyze daily precipitation. Comparing data from the continental US from 2015 and 2021, you will observe patterns in the data to determine if there are distinct observable differences.

First, I will introduce you to Xarray, an open source project that interoperates with NumPy, SciPy, and matplotlib, extending beyond NumPy ndarrays and pandas dataframes.

As with the preceding chapters, after you complete an introduction to a package and its supporting documentation, I strongly encourage you to experiment with different datasets and applications of Python packages and libraries.

Downloading Your Data

First, navigate to Gridded Climate Datasets: Precipitation (*https://oreil.ly/NhTn2*) and choose the Climate Prediction Center (CPC) Global Precipitation dataset (*https://oreil.ly/M8ynH*). Weather data for a given latitude/longitude is returned for the grid cell aligned with the requested lat/long.

Download the years of interest, 2015 and 2021, and upload the files to either your Google drive or directly to your computer. Figure 10-1 shows the folder and file hierarchy.

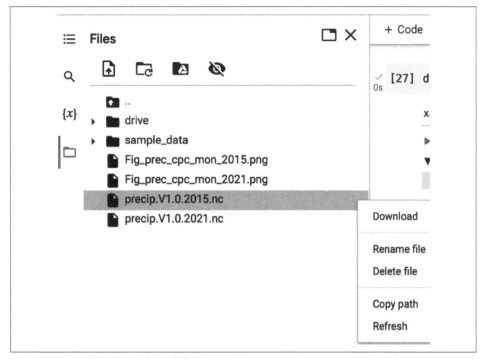

Figure 10-1. Files in Google Colab

Working in Xarray

Xarray is a Python library that hosts many dependencies that should by now be familiar to you, such as NumPy and pandas, as well as a few optional dependencies that you'll need to work with the CPC Global Precipitation dataset. Originally developed by the Climate Corporation, Xarray (*https://oreil.ly/NjJmL*) has become a useful open source resource for analyzing climate change data files for plotting and analysis. It's useful for exploring weather, water, and climate extremes and their impact.

The NOAA Physical Sciences Laboratory relies on the Network Common Data Form (netCDF) format of Xarray, which is an interface for array-oriented data containing dimensions, variables, and attributes on top of NumPy arrays. The dimensions are often time and latitude or longitude, so this format is directly applicable to spatial observation and analysis.

You will download daily precipitation data and apply functions to it, including `groupby` (for grouping), `concat` (to combine files, also called *concatenation*), and `sel` & `isel` (to select data for specific dates or particular locations). Additionally, you will learn how to handle leap years. You'll save the desired outputs as netCDF files.

Gridded data combines point data (such as data from an individual weather station) and other data sources and maintains a spatially and temporally consistent method to account for factors like temperature changes and precipitation caused by location or elevation. The weather data is provided from a data distribution containing more than 2,500 monthly gridded data points, representing the entire globe.

Xarray's data structures are *N-dimensional*, meaning that they have a variable number of dimensions. This makes Xarray suitable for dealing with multidimensional scientific data. Because it uses dimension names instead of axis labels (dim='time' instead of axis=0), these arrays are more manageable than arrays in NumPy ndarrays. With Xarray, you don't need to keep track of the order of dimensions or insert placeholder dimensions of size 1 to align arrays.

You will work with a few of the available functions to analyze geospatial data. Running `open_mfdataset` opens multiple files at one time. You will use this to compare different timepoints. You are going to run these examples in Google Colab (*https://oreil.ly/J8wam*). I always connect to my Google Drive, as I keep most of the datasets on the cloud and not on my local computer. It is simple to connect your drive to Colab and select the directory where you are hosting the dataset:

```
from google.colab import drive
drive.mount('/content/drive')
```

 Not everyone has a Google Drive. You can also access your data by simply uploading it to Google Colab. I actually upgraded to Google Colab Pro to improve runtime performance. Connecting to your Google Drive is usually the best option, but your mileage may vary.

You will need to establish the connection by selecting and approving it. Mounting the drive will take a little time, but then the files will show up in your available files.

Google Colab has a preinstalled version of Xarray, but I suggest a `pip install` to make certain the dependencies are all included:

```
!pip install xarray[complete]
```

I make it a habit to run installs separately but bundle the import functions; this allows me to isolate and address any errors.

Let's take a look at the less familiar modules you will be importing. The `glob` module (short for "global") returns the files or folders that you specify. This also applies to paths inside directories/files and subdirectories/subfiles. You should recognize matplotlib as our plotting library, built on NumPy, and `urllib.request` as a module for retrieving URLs. You can download data to upload to Colab or, if your data is available as a URL, add the import to your code:

```
import glob
import matplotlib.pyplot as plt
import urllib.request
import xarray as xr
```

Combining Your 2015 and 2021 Datasets

The Xarray dataset is a container of labeled arrays. It is similar to a dataframe, except it is multidimensional and aligns with the netCDF dataset representation, shown in Figure 10-2.

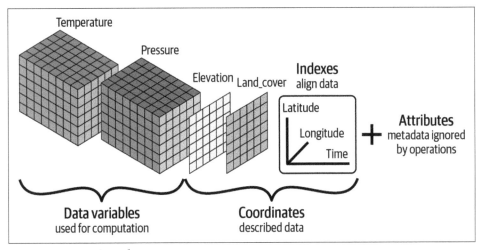

Figure 10-2. NetCDF data array

Enter the code to create your two variables, ds2015 for 2015 and ds2021 for 2021:

```
ds2015 = xr.open_dataset('/content/precip.V1.0.2015.nc')
```

```
ds2021 = xr.open_dataset('/content/precip.V1.0.2021.nc')
```

Figure 10-3 shows the key properties of the Xarray dataset, including dimensions, coordinates, data variables, and attributes.

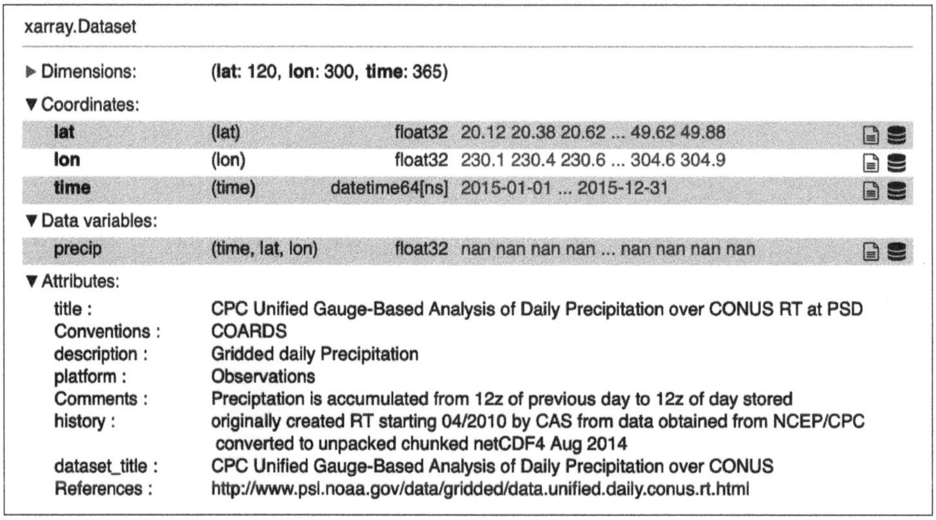

```
xarray.Dataset
▶ Dimensions:          (lat: 120, lon: 300, time: 365)
▼ Coordinates:
    lat            (lat)                float32  20.12 20.38 20.62 ... 49.62 49.88       📄 🗄
    lon            (lon)                float32  230.1 230.4 230.6 ... 304.6 304.9       📄 🗄
    time           (time)         datetime64[ns]  2015-01-01 ... 2015-12-31             📄 🗄
▼ Data variables:
    precip         (time, lat, lon)     float32  nan nan nan nan ... nan nan nan nan     📄 🗄
▼ Attributes:
    title :            CPC Unified Gauge-Based Analysis of Daily Precipitation over CONUS RT at PSD
    Conventions :      COARDS
    description :      Gridded daily Precipitation
    platform :         Observations
    Comments :         Preciptation is accumulated from 12z of previous day to 12z of day stored
    history :          originally created RT starting 04/2010 by CAS from data obtained from NCEP/CPC
                         converted to unpacked chunked netCDF4 Aug 2014
    dataset_title :    CPC Unified Gauge-Based Analysis of Daily Precipitation over CONUS
    References :       http://www.psl.noaa.gov/data/gridded/data.unified.daily.conus.rt.html
```

Figure 10-3. Dataset properties in Xarray

Click the database icon (it looks like stacked discs) to see more details. You will see the cell expand, as shown in Figure 10-4. The expanded metadata reveals the array and additional information.

Next, you are going to concatenate (*https://oreil.ly/E60L2*) (join) Xarray objects along the time dimension:

```
ds2015_2021 = xr.concat([ds2015,ds2021], dim='time')
ds2015_2021
```

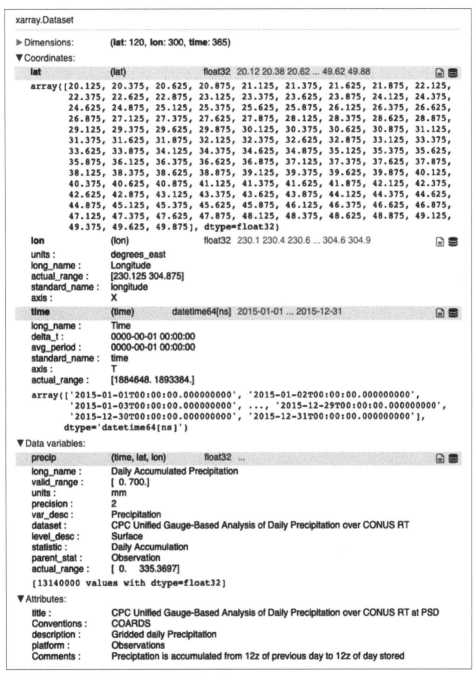

Figure 10-4. Expanded dataset details

The output (Figure 10-5) will confirm that the datasets have been combined. The time coordinate now lists both years that you selected.

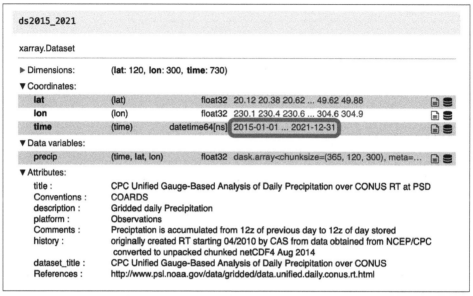

Figure 10-5. Time coordinates concatenated

The function `dataarray.groupby` returns an object for group operations. You will be grouping operations by month (`time.month` in the code). There is a `dataarray` object for a daily dataset that spans a few years. This object has one variable and three dimensions: latitude, longitude, and time (daily). You can generate a graphic for both 2015 and 2021 (Figure 10-6).

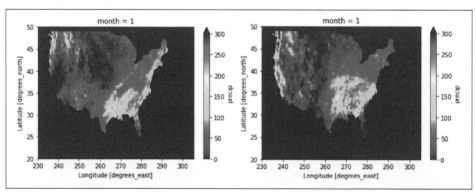

Figure 10-6. Monthly precipitation in the continental US in 2015 (left) and 2021 (right)

Generating the Images

To generate the plot for your 2015 image, state the lower bound by the first item (0), the default for the upper bound, and the step size default. Monthly precipitation is an object to be sliced, as indicated by mon.precip. We are operating across the three dimensions:

```
ds2015_mon = ds2015.groupby('time.month').sum()
ds2015_mon.precip[0,:,:].plot(cmap='jet', vmax=300)
```

You can use the same method to generate a map for 2021. The output for both is in Figure 10-6, and the different patterns of precipitation are visible.

Let's expand beyond a single month to compare different years. Python has a calendar module that will display an entire year for comparison. Follow this code to iterate over each month in a calendar year, using matplotlib and the calendar module:

```
import calendar
```

Now, you will apply the sum along the time dimension. The landmask you are creating will "mask out" certain observations instead of returning NaN values:

```
landmask = ds2015.precip.sum(dim='time')>0
```

Matplotlib allows you to format and plot a series of maps. There are defaults for these parameters, but when you want to customize them, you can provide a width and height in inches for the image (figsize) and a background color (facecolor). The parameter subplots_adjust allows you to adjust the position of the subplot or axes. Let's do 2015 first:

```
fig = plt.figure(figsize=[12,8], facecolor='w')
plt.subplots_adjust(bottom=0.15, top=0.96, left=0.04, right=0.99,
                    wspace=0.2, hspace=0.27)
nrows = 3
ncols = 4
for i in range(1, 13):
#the python data index starts at 0, but the subplot starts at 1.
plt.subplot(nrows, ncols, i)
    dataplot = ds2015_mon.precip[i-1, :].where(landmask)
    p = plt.pcolormesh(ds2015_mon.lon, ds2015_mon.lat, dataplot,
                vmax = 400, vmin = 0, cmap = 'nipy_spectral_r',
                )
    plt.xlim([233,295])
    plt.ylim([25,50])
    plt.title(calendar.month_name[dataplot.month.values], fontsize = 13,
            fontweight = 'bold', color = 'b')
    plt.xticks(fontsize = 11)
    plt.yticks(fontsize = 11)
    if i % ncols == 1: # Add ylabel for the very left subplots
        plt.ylabel('Latitude', fontsize = 11, fontweight = 'bold')
    if i > ncols*(nrows-1): # Add xlabel for the bottom row subplots
```

```
plt.xlabel('Longitude', fontsize = 11, fontweight = 'bold')

# Add a colorbar at the bottom:
cax = fig.add_axes([0.25, 0.06, 0.5, 0.018])
cb = plt.colorbar(cax=cax, orientation='horizontal', extend = 'max',)
cb.ax.tick_params(labelsize=11)
cb.set_label(label='Precipitation (mm)', color = 'k', size=14)

# Now we can save a high resolution (300dpi) version of the figure:
plt.savefig('Fig_prec_cpc_mon_2015.png', format = 'png', dpi = 300)
```

Experiment with different values to see how the image changes. The edges of the subplots are nudged along with the amount of width space (wspace) or height space (hspace).

You can explore different plot types in the matplotlib (*https://oreil.ly/7Nrc9*) usage guide. The output is shown in Figure 10-7.

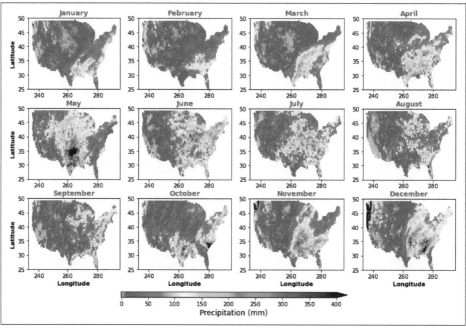

Figure 10-7. Yearly precipitation in the continental US by month, 2015, in Xarray

Next, update the code cell to run the 2021 data (Figure 10-8) and visually compare the difference in precipitation between the two years.

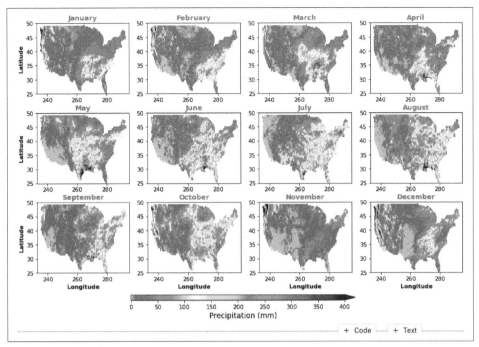

Figure 10-8. Yearly precipitation, 2021

More Exploration

There are additional queries you can try if you want to see the data underlying the imagery. For instance, `ds2021.precip.dims` will list the dimensions in the data, `ds2021.precip.attrs` will list attributes, and `ds2021.precip.data` will show the daily precipitation.

Precipitation data might be more meaningful if grouped seasonally, so let's try that:

```
ds2021.groupby("time.season")
DatasetGroupBy, grouped over 'season'
4 groups with labels 'DJF', 'JJA', 'MAM', 'SON'.
```

The seasons are named by the months they contain: DJF for winter, JJA for summer, MAM for spring, and SON for fall. Let's order them in an intuitive manner:

```
seasonal_mean = seasonal_mean.reindex(season=["DJF", "MAM", "JJA", "SON"])
seasonal_mean
```

The seasons are now displayed chronologically, as they occur in a single year. Visualizing precipitation levels allows you to observe how precipitation varies by season and location (Figure 10-9).

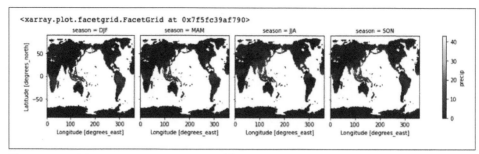

Figure 10-9. Mean precipitation arranged by season

Observations of precipitation levels demonstrate patterns in the data. You may have noticed that temperature datasets are also available in gridded format. Now that you have been introduced to Xarray and how to work with multidimensional arrays, try exploring maximum temperature, minimum temperature, and average temperature to expand your insights. What can you learn by plotting this data as well?

Example 2: Deforestation and Carbon Emissions in the Amazon Rain Forest Using WTSS Series

The Amazon rain forest is shrinking. Roadway construction in the forest, contact between humans and animals, and contact between visitors and the Indigenous peoples who live in the rain forest all contribute to ongoing deforestation and forest degradation. *Degradation* refers to a decline in forest density that does not quite meet the level of deforestation—typically, when tree cover is diminished between 10% and 30% and the land is being converted to alternative uses, such as timber harvesting or agriculture. When the forest cover is disturbed, *biodiversity*—the Amazon's mixture of living things and ecosystems that exists nowhere else on earth—is threatened. Deforestation also has an impact on carbon emissions.

Web Time Series Service (WTSS) is a web service that processes time-series data gathered by remote sensing imagery.[1] Again, you'll be working with a 3D array with spatial (latitude and longitude) and temporal references. WTSS is a Python client library, so you can obtain a list of actual values recorded at a specific location over time.

---

1 To learn more, I recommend this paper, which introduces WTSS to create a time series from remote sensing data in the Brazilian rainforest: Vinhas, L., Queiroz, G. R., Ferreira, K. R., and Camara, G. 2017. "Web Services for Big Earth Observation Data." *Revista Brasileira de Cartografia* 69: 5. English translation (*https://oreil.ly/rbckx*).

Setup

In this exercise, you will explore data from a research project of Brazil's National Institute for Space Research (INPE), Brazil Data Cube (BDC) (*https://oreil.ly/wZ1DD*). This data cube is a temporal composite from CBERS-4 and surface reflectance data from Advanced Wide Field Imager (AWFI), an instrument that captures high-resolution land and vegetation imagery.

Obtaining the data

Begin by registering at the website. You will need to create a profile, then generate and save an access token using the menu pictured in Figure 10-10.

Figure 10-10. BDC access token

The BDC project collects continuous data on land cover (an example is shown in Figure 10-11).[2] To identify an area of interest, select the cube stack CBERS-4-AWFI within the CBB4_64_16D_STK-1 collection. The information icon ("i") will provide background information.

---

2 Earth Observation Data Cubes for Brazil: Requirements, Methodology and Products, Earth Observation and Geoinformatics Division, National Institute for Space Research (INPE), Avenida dos Astronautas, 1758, Jardim da Granja, Sao Jose dos Campos, SP 12227-010, Brazil.

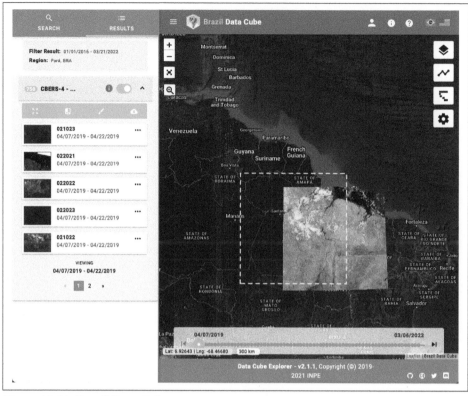

Figure 10-11. Brazil Data Cube

You'll investigate if the vegetation density of the Amazon has changed over time by computing the NDVI, which you learned about in Chapter 6. Recall that the NDVI calculates the difference between near-infrared (which is reflected by vegetation) and red light (absorbed by vegetation). These bands highlight vegetated areas and allow you to observe the density of vegetative growth. You will select a location in the Amazon rain forest within a region of dense vegetation. Enter the coordinates: latitude = –16.350000 and longitude = –56.666668.

As you explore BDC on your own after finishing this exercise, try entering latitude and longitude coordinates of your own choosing. As an alternative, you can also use the bounding box (bbox) information to locate areas of interest.

Creating your environment

Open a Jupyter Notebook and install the libraries using `pip install`. (I know, I know—I typically advocate for Conda installs, but the WTSS is only stable in pip. Often, we must adapt!)

```
!pip install wtss
from wtss import WTSS
```

To access WTSS, use the link in the following code and the BDC access token you generated when creating your account:

```
service = WTSS('https://brazildatacube.dpi.inpe.br/',
    access_token='Your access token')
```

Next, run:

```
service.coverages
```

This outputs a list of services available at the link:

```
['LANDSAT-MOZ_30_1M_STK-1',
 'MOD13Q1-6',
 'LC8_30_6M_MEDSTK-1',
 'MYD13Q1-6',
 'S2-SEN2COR_10_16D_STK-1',
 'CB4_64_16D_STK-1',
 'LC8_30_16D_STK-1',
 'CB4MUX_20_1M_STK-1']
```

Creating Your Map

Select the CBERS coverage for the code cell:

```
cbers4_coverage = service['CB4_64_16D_STK-1']
cbers4_coverage
```

The output (Figure 10-12) provides the attributes of the CB4_64 cube.

Description: This datacube was generated with all available surface reflectance images from CB4_64 cube. The data is provided with 64 meters of spatial resolution, reprojected and cropped to BDC_LG grid, considering a temporal compositing function of 16 days using the best pixel approach (Stack).

Attributes

name	common name	description	datatype	valid range	scale	nodata
NDVI	ndvi	Normalized Difference Vegetation Index	int16	{'min': -10000.0, 'max': 10000.0}	0.0001	-9999.0
BAND16	nir	Band 16 (nir)	int16	{'min': 0.0, 'max': 10000.0}	0.0001	-9999.0
CMASK	quality	Cloud Mask	uint8	{'min': 0.0, 'max': 255.0}	1.0	0.0
BAND13	blue	Band 13 (blue)	int16	{'min': 0.0, 'max': 10000.0}	0.0001	-9999.0
BAND14	green	Band 14 (green)	int16	{'min': 0.0, 'max': 10000.0}	0.0001	-9999.0
BAND15	red	Band 15 (red)	int16	{'min': 0.0, 'max': 10000.0}	0.0001	-9999.0
CLEAROB	ClearOb	Clear Observation Count	uint8	{'min': 1.0, 'max': 255.0}	1.0	0.0
PROVENANCE	Provenance	Provenance value Day of Year	int16	{'min': 1.0, 'max': 366.0}	1.0	-1.0
TOTALOB	TotalOb	Total Observation Count	uint8	{'min': 1.0, 'max': 255.0}	1.0	0.0
EVI	evi	Enhanced Vegetation Index	int16	{'min': -10000.0, 'max': 10000.0}	0.0001	-9999.0

Figure 10-12. CB4_64 datacube with description

Notice how, when you assign variables, you use the name *exactly* as it appears in the table. We are interested in the near-infrared and red bands, so let's assign variables for those:

```
red_band = 'BAND15'
nir_band = 'BAND16'
```

The time-series data for the location and dates listed is retrieved by the `ts` method:

```
time_series = cbers4_coverage.ts(attributes=(red_band, nir_band),
                                 latitude=-16.350000,
                                 longitude= -56.666668 ,
                                 start_date="2016-01-01",
                                 end_date="2022-03-21")
```

Plot the static visualization of the time series with the `plot` method:

```
time_series.plot()
```

The output is pictured in Figure 10-13.

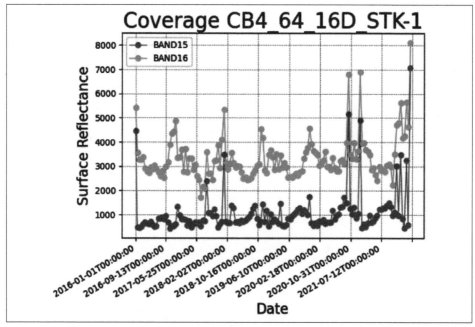

Figure 10-13. Static visualization of CB4_64_16D_STK-1 data cube time series

Analysis

As you can see in Figure 10-13, the NDVI in this area of the Amazonian rain forest trends upward between 2016 and 2022. This could suggest an increase in atmospheric carbon dioxide (CO_2) levels, although precipitation and temperature are more important for explaining interannual NDVI variability. Indeed, ecosystem models suggest that most of the observed increase in the seasonal amplitude of atmospheric CO_2, indicated by increasing plant growth over recent decades, is mostly due to CO_2 fertilization, with climate and land-use changes playing secondary roles. We can't claim anything definitively based on this data, but the ability to capture surface reflectance and observe NDVI can help in generating hypotheses.[3]

Refinements

Let's look at a few more ways to improve this map.

Making your map interactive

Next, try creating an interactive display of your findings, using the pandas dataframe:

```
cbers_df = pd.DataFrame({ 'BAND15': time_series.BAND15,
'BAND16': time_series.BAND16 },
                        index = pd.to_datetime(time_series.timeline))
cbers_df
```

You can hover in the notebook to watch the information display (Figure 10-14). Select different bands in different years for additional information.

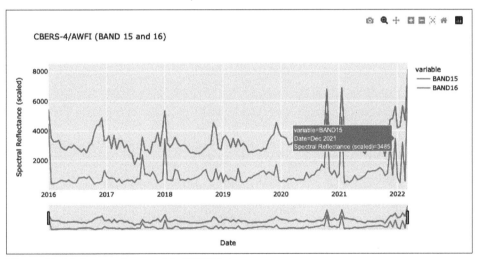

Figure 10-14. Interactive graphic CBERS-4

3 Ito, Akihiko. 2019. "Disequilibrium of Terrestrial Ecosystem CO_2 Budget Caused by Disturbance-Induced Emissions and Non-CO_2 Carbon Export Flows: A Global Model Assessment." *Earth System Dynamics* 10: 685–709. *https://doi.org/10.5194/esd-10-685-2019.*

`px.line()` is calling a plotly function. The `dataframe` parameter allows you to plot data. The x parameter specifies the variable on the *x*-axis, and the y parameter does the same for the *y*-axis. The title is described as well:

```
fig = px.line(cbers_df, x=cbers_df.index, y=['BAND15', 'BAND16'],
title='CBERS-4/AWFI (BAND 15 and 16)', labels={
    'index': 'Date',
    'value': 'Spectral Reflectance (scaled)'
})

fig.update_xaxes(rangeslider_visible=True)
fig.show()
```

Reducing cloud coverage with masking

Satellite images observed in time series often display noise or distortion from clouds, which can change the spectral behavior of the regions being analyzed. Clouds and their shadows can decrease reflectance.

Interpolation is one way of replacing pixel values or points that have been distorted by cloud cover. The process of interpolation constructs new data points within the range of discrete values of known data points. Data cubes contain *masks* that capture the influence of cloud cover, pixel by pixel.

The Jupyter Notebook contains additional exercises working with masking and interpolation using BDC.

Remote sensing technology relies on cloud masking to process data imagery and improve its quality. Cloud masks let users identify cloudy and cloud-free pixels by analyzing the percentage of cloud in the mask.

The code in this section will walk you through first detecting clouds, then obtaining a mask to "hide" any low-quality data and prevent it from skewing the results.

When you run the code, it will output metadata with one of three values:

- A value of 0 means the pixel contains no data.
- A value of 127 means the pixel is clear, with no clouds.
- A value of 255 means the pixel is obscured by clouds.

Let's look at a slightly different location:

```
cb4_timeseries_cmask = cbers4_coverage.ts(

    attributes = ('CMASK'),

    latitude = -12.0,
    longitude = -53.989,

    start_date = "2017-01-01",
    end_date = "2021-12-31"
)

cb4_timeseries_cmask
```

This gives the output:

```
CMASK: [127.0, 127.0, 127.0, 127.0, 127.0, 127.0, 127.0, 127.0, 127.0, 127.0,
127.0, 127.0, 127.0, 127.0, 127.0, 127.0, 127.0, 127.0, 127.0, 127.0, 127.0,
127.0, 127.0, 255.0, 127.0, 127.0, 127.0, 255.0, 127.0, 127.0, 127.0, 127.0,
127.0, 127.0, 127.0, 127.0, 127.0, 127.0, 127.0, 127.0, 127.0, 127.0, 127.0,
127.0, 127.0, 127.0, 127.0, 127.0, 127.0, 127.0, 127.0, 127.0, 127.0, 127.0,
127.0, 127.0, 127.0, 127.0, 127.0, 127.0, 127.0, 127.0, 127.0, 127.0, 127.0,
127.0, 127.0, 127.0, 127.0, 127.0, 127.0, 127.0, 255.0, 127.0, 127.0, 127.0,
127.0, 127.0, 127.0, 127.0, 127.0, 127.0, 127.0, 127.0, 127.0, 127.0, 127.0,
127.0, 127.0, 127.0, 127.0, 127.0, 127.0, 127.0, 127.0, 127.0, 255.0]
```

You can see that most of the pixels are clear, but a few cloud observations record a value of 255.0. The set function displays the values as well:

```
set(cb4_timeseries_cmask.values('CMASK'))
```

The output is {127.0, 255.0}.

The following code will display cloud observations over the time selected in the range:

```
cb4_timeseries = cbers4_coverage.ts(

    attributes = ('NDVI', 'CMASK'),

    latitude = -12.0,
    longitude = -53.989,

    start_date = "2017-01-01",
    end_date = "2019-12-31"

)
```

The timeline data is stored as a list of dates, but you will need to transform it to a datetime object. To do that, run the following code:

```
ndvi_timeline = pd.to_datetime(cb4_timeseries.timeline)
ndvi_timeline
```

Now, you'll store the transformed data in the variables ndvi_data and cmask_data:

```
import numpy as np

ndvi_data = np.array(cb4_timeseries.NDVI)
ndvi_data

cmask_data = np.array(cb4_timeseries.CMASK)
cmask_data
```

The next step is to remove the clouds. The NumPy array (np) will convert all 127 values (no clouds) to 1, and the 255 (cloudy) values will now show as NaN (not a number):

```
cmask_data = np.where(cmask_data == 255, np.nan, 1)
cmask_data
```

Now multiply the NaN values by the NDVI array to interpolate the time-series data:

```
ndvi_data * cmask_data
```

This creates a new dataframe for the NaN values:

```
ndvi_masked_data = pd.DataFrame({ 'data': ndvi_data * cmask_data },
index = pd.to_datetime(ndvi_timeline))
ndvi_masked_data[ndvi_masked_data['data'].isna()]
```

And sure enough, when you run the code, there they are in the output (Table 10-1).

Table 10-1. NaN values in interpolated cloud mask data

2018-01-01	NaN
2018-03-06	NaN
2020-02-18	NaN
2021-03-06	NaN

Now we can pull out all of the NaN data and view the interpolated data:

```
ndvi_masked_data_interpolated = ndvi_masked_data.interpolate()
ndvi_masked_data_interpolated[ndvi_masked_data_interpolated['data'].isna()]
```

Let's visualize the interpolated data! Run this code:

```
plt.figure(dpi = 120)

plt.plot(ndvi_data, color='gray', linestyle='dashed', label = 'Original')
plt.plot(ndvi_masked_data_interpolated['data'].values, color='blue',
label = 'Interpolated')

plt.title('Comparison of Time Series with and without interpolation')
plt.legend()
plt.grid(True)
plt.show()
```

Your output should look like Figure 10-15. Removing the cloud observations yields unobstructed values of the pixels and the surface reflectance to more accurately depict the vegetation density.

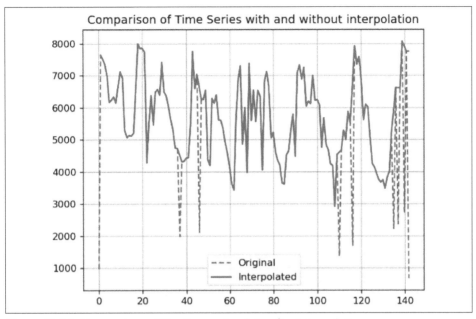

Figure 10-15. Comparison of the time-series data after interpolating

This exercise showed you how to use WTSS to access and analyze remote sensing data with Python and APIs and introduced you to the BDC resource. You also saw that data analysis isn't just for backing up your arguments; in fact, exploring data with these tools can be an important part of the process of *generating* a hypothesis and can lead your inquiries to new and unexpected places.

Example 3: Modeling and Forecasting Deforestation in Guadeloupe with Forest at Risk

Beyond the benefit of modeling the spatial probability of deforestation, this final section is a great example of modeling large georeferenced raster files. We will be working with a Python package that analyzes deforestation: Forest at Risk (FAR) (*https://oreil.ly/2MLdk*). The functions in this package process blocks of data modeled over large geographic scales at a high resolution.

The area of interest is Guadeloupe, an archipelago and overseas department of France located in the eastern Caribbean Sea. According to Global Forest Watch, more than 240 acres of Guadeloupe's humid primary forest have been lost in 2021 alone. Tree-cover loss was responsible for about 1.07 metric tons (2,358.95 pounds) of CO_2 emissions.

The goal of working with FAR is to model Guadeloupe's deforestation spatially, predict the risk of deforestation based on a set of variables, and forecast future forest density. The spatial variables you'll be working with include topography (altitude, slope, and aspect), accessibility (distance to roads, towns, and the forest edge), and distance to previous areas of deforestation or protected areas.

Any statistical model class with a `.predict()` method can potentially be used with the function `forestatrisk.predict_raster()` to predict the spatial risk of defor- estation. Performing computations on arrays using NumPy and GDAL increases efficiency without the need for high computational power. You'll use this data to define the geographic area and to measure deforestation, forest degradation, and forest regrowth. Let's see if we can visualize the change in surface reflectance and underlying tree cover by relying on geospatial tools.

We'll be using data from International Forest Resources and Carbon Emissions (IFORCE), part of the European Commission's Joint Research Center, which is responsible for independent scientific and technical support. This data measures Landsat time-series forest-cover change in tropical moist forests (TMFs). It was gathered from seasonal rain forests at increasing distance from the equator over 30 years, from 1990 to 2020.

Setup

Although I am working with Google Colab in this example, there were challenges with reproducibility, and I found creating a Conda environment to be a more stable option. The dashboard in Figure 10-16 shows the output files that will not show up in your notebook automatically. Select them from the file hierarchy in Colab or from your Jupyter Notebook file structure.

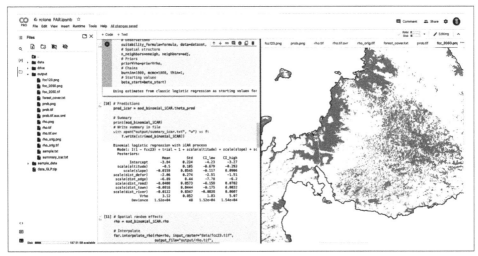

Figure 10-16. Opening the FAR dataset in Google Colab Pro

Depending on your individual setup, Google Colab usually runs quite well. I default to Colab when instructing because it is straightforward in highlighting the folder structure within your Google Drive (or the import paths, if you choose to upload files). There are occasions when you might not be able to get this to work. In those instances, I have found creating an environment in the terminal to be the solution.

Creating your environment

I am using Python 3.10, which as of this writing is the current version, but you'll enter your own version when you run the following code. Everything should work, but note that when you build environments and adjust versioning, the details might need a little fine-tuning. (You are far enough along in your Python journey to be familiar with a few of these options, but return to the preface if you need a reminder.)

Now you can create your environment using Conda:

```
conda create --name conda-far -c conda-forge python=3.7 gdal numpy matplotlib
pandas patsy pip statsmodels earthengine-api --yes
conda activate conda-far
pip install pywdpa sklearn # Packages not available with conda
pip install forestatrisk # For PyPI version
conda install -c conda-forge python-dotenv rclone
```

If you get an error message that forestatrisk is not recognized, you can create a kernel inside your FAR environment in the terminal:

```
conda install ipykernel
python -m ipykernel install --user --name myenv
--display-name "FAR"
```

Enter **jupyter notebook** and select the FAR kernel. Your package should now run.

Downloading and importing packages

Importing the necessary files and packages is the first step. The os module allows you to interact with folders on your operating system. The shutil and copy2 packages preserve the metadata of the files you are importing, including permissions. (There are limits to this when working on MacOS, so consult the Python documentation (*https://oreil.ly/cJTAA*).) The urllib package collects several modules for opening and reading URLs. You'll also be working with ZIP files, and the ZipFile module provides the necessary tools.

Go ahead and import your packages:

```
# Imports
import os
from shutil import copy2
import urllib.request
from zipfile import ZipFile

import forestatrisk as far

import numpy as np
import matplotlib.pyplot as plt
import pandas as pd
from patsy import dmatrices
import pickle
from sklearn.linear_model import LogisticRegression
from sklearn.ensemble import RandomForestClassifier
from sklearn.metrics import log_loss
```

Downloading and importing the data

Next you'll create an output directory, which is especially convenient when working in a notebook environment. When you run your scripts, the output files will be in this designated location, including the figures you are generating:

```
# Make output directory
far.make_dir("output")
```

The ZIP file *data_GLP.zip* is located at the link in the following code block, and z.extractall will unzip the file into your Colab or Jupyter Notebook directory, as shown in Figure 10-13:

```
# Source of the data
url = "https://github.com/ghislainv/forestatrisk/raw/master/docsrc/notebooks/
data_GLP.zip"

if os.path.exists("data_GLP.zip") is False:
    urllib.request.urlretrieve(url, "data_GLP.zip")

with ZipFile("data_GLP.zip", "r") as z:
    z.extractall("data")
```

The *.zip* file contains a selection of environmental variables and maps from across a timespan: all of the data you will need for this brief walk through the FAR tutorial, in which you will model the spatial probability of deforestation in a specific geography.

As you download the data (*https://oreil.ly/lkzTB*), take a few moments to explore the details regarding symbology, values, and labels on the IFORCE (*https://oreil.ly/wtHAi*) site.

Plotting the Data

Run the following code to plot the data:

```
# Plot forest
fig_fcc23 = far.plot.fcc(
    input_fcc_raster="data/fcc23.tif",
    maxpixels=1e8,
    output_file="output/fcc23.png",
    borders="data/ctry_PROJ.shp",
    linewidth=0.3, dpi=500)
```

When you run the code cell, this code will generate a map as output, but it will not populate in the notebook automatically. To find the output files in Google Colab, use the dashboard to navigate through the folder hierarchy to your output folder. In a Jupyter Notebook, review the files listed in the tab at the top left corner (Figure 10-17). The data files you downloaded will also be visible.

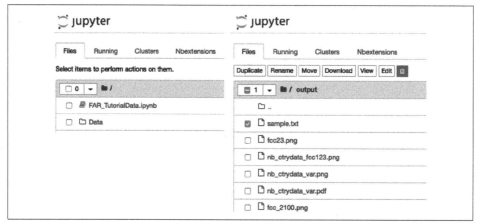

Figure 10-17. Locating your output in the Jupyter Notebook file structure

When you plot the data (Figure 10-18), the forest appears as green and the deforested areas as red. Can you see the perimeter where deforestation is more likely?

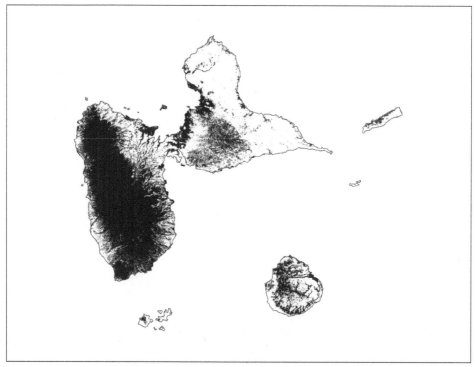

Figure 10-18. Plotting FAR data on forest cover change in Guadeloupe as a GeoTiff raster file

Sampling the Data

Next, it's time to sample your data. You'll use the function `far.sample`. The help documentation, which you can access by running `help(far.sample)`, tells us that this function "randomly draws spatial points in deforested and forested areas and… extract[s] environmental variable values for each spatial point." In other words, it samples 10,000 randomly selected pixel centers (points) in the areas identified as either deforested or as remaining forest.

Let's take a moment to look at the arguments this function takes. First, `seed` tells the function to reproduce the random sampling data from the previous step. Next, the spatial cells for each sample are grouped by the `csize` argument. Each spatial cell is attributed to a parameter, so they need to be of a sufficient size to ease the burden of memory and large datasets. These grouped observations are estimates of spatial autocorrelation in deforestation.

Remember Tobler's First Law, introduced in Chapter 1: "Everything is related to everything else, but near things are more related than distant things." *Spatial autocorrelation* is a measure of spatial variation within adjacent observations where it is possible for near things to not be similar! Adjacent observations can have similar values (positive spatial autocorrelation) or have contrasting values (negative autocorrelation). The presence or absence of a pixel value between measurements has important implications for spatial statistics. There are many reasons why this might be the case, but one could be that a small sample from a much larger dataset simply isn't representative of the larger sample.

In the following code cell, you'll also see the argument `var_dir`. This points to the directory where all of the GeoTIFF files are located and to *fcc23.tif* as a forest-cover-change raster file. Once more, pay attention to the output file and notice it appearing in your directory:

```
# Sample points
dataset = far.sample(nsamp=10000, adapt=True, seed=1234, csize=10,
                     var_dir="data",
                     input_forest_raster="fcc23.tif",
                     output_file="output/sample.txt",
                     blk_rows=0)
```

I'm showing you the output here to demonstrate how the calculations are conducted:

```
Sample 2x 10000 pixels (deforested vs. forest)
Divide region in 168 blocks
Compute number of deforested and forest pixels per block
100%
Draw blocks at random
Draw pixels at random in blocks
100%
Compute center of pixel coordinates
```

```
Compute number of 10 x 10 km spatial cells
... 99 cells (9 x 11)
Identify cell number from XY coordinates
Make virtual raster with variables as raster bands
Extract raster values for selected pixels
100%
Export results to file output/sample.txt
```

The following code removes "not available" (NA) values from the dataset, then computes the number of neighbors of each spatial cell. It outputs a pandas dataframe:

```
# Remove NA from dataset
dataset = dataset.dropna(axis=0)
# Set number of trials to one for far.model_binomial_iCAR()
dataset["trial"] = 1
# Print the first five rows
print(dataset.head(5))
```

If you do not remove the NAs, you won't be able to model the data in the next steps—recall that the logistic regression model is reporting on the presence or absence of a pixel value between measurements.

The output shown here is a sample of the first five rows. The variables represent the spatial explanatory variables used to model the impact of deforestation and include topography represented by altitude and slope; accessibility, measured as distance to nearest road, town, river, and forest edge; deforestation history, calculated as distance to past deforestation; and land conservation status as a protected area (pa):

```
   altitude  dist_defor  dist_edge  dist_river  dist_road  dist_town  fcc23  \
0      30.0       642.0       30.0      8448.0     1485.0     6364.0    0.0
1      37.0       765.0       30.0      8583.0     1697.0     6576.0    0.0
2      78.0       216.0       30.0      7722.0      949.0     5743.0    0.0
3      80.0       277.0       30.0      8168.0     1172.0     6047.0    0.0
4      46.0        30.0       30.0      6179.0      541.0     6690.0    0.0

    pa  slope          X          Y              cell  trial
0  0.0    8.0 -6842295.0  1851975.0  4.0     1
1  0.0    7.0 -6842235.0  1852095.0  4.0     1
2  0.0    5.0 -6842535.0  1851195.0  4.0     1
3  0.0    2.0 -6842445.0  1851615.0  4.0     1
4  0.0    1.0 -6840465.0  1849755.0  4.0     1
```

Sample-size calculations for forest cover change the map—the pixel value is set to 1 when there is forest and to 0 when there is no forest:

```
# Sample size
ndefor = sum(dataset.fcc23 == 0)
nfor = sum(dataset.fcc23 == 1)
with open("output/sample_size.csv", "w") as f:
    f.write("var, n\n")
    f.write("ndefor, " + str(ndefor) + "\n")
```

```
        f.write("nfor, " + str(nfor) + "\n")
print("ndefor = {}, nfor = {}".format(ndefor, nfor))
```

Correlation Plots

The next step is to correlate all this data. This means actually comparing the locations of forested areas with the other variables, such as roads, rivers, towns, and other deforested areas, to plot how likely it is that the point in question will be deforested.

Run the code to correlate and then plot the result:

```
# Correlation formula
formula_corr = "fcc23 ~ dist_road + dist_town + dist_river + \
dist_defor + dist_edge + altitude + slope - 1"

# Output file
of = "output/correlation.pdf"
# Data
y, data = dmatrices(formula_corr, data=dataset,
                    return_type="dataframe")
# Plots
figs = far.plot.correlation(
    y=y, data=data,
    plots_per_page=3,
    figsize=(7, 8),
    dpi=80,
    output_file=of)
```

This outputs the *correlation plots* shown in Figure 10-19.

You can read the probability of deforestation by looking first at the distribution of the data. Notice the shape of the probability curve. Look at how far the distance a pixel is from a deforested state. In Figure 10-16, you can see that the bulk of pixels in the sample are relatively close to a deforested area already—and, as you know, as the distance increases, the probability of deforestation declines.

Looking at the parameter estimates in Figure 10-19, you can see that the probability of deforestation decreases with altitude, slope, distance to past deforestation, and forest edge. The distances to road, town, and river all "cross zero," meaning that the values are not significantly different from zero. The confidence interval range includes zero.

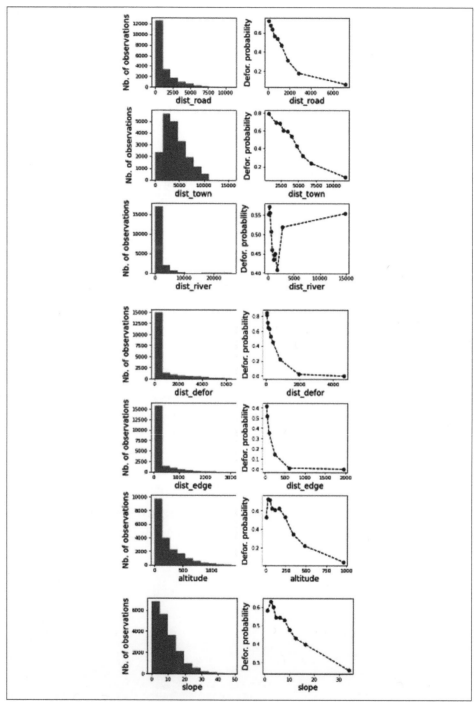

Figure 10-19. Correlation plots of selected variables

Modeling the Probability of Deforestation with the iCAR Model

You won't need to understand the underlying statistics to be able to render the output. The statistical model `model_binomial_iCAR` estimates the probability of deforestation by examining each pixel and evaluating a set of environmental variables like altitude, distance to edge of forest, distance to road, and distance to the nearest towns. These variables tell us how accessible the forest is to humans. In simple terms, we need this data because fixed environmental variables don't completely explain the process of deforestation, at least not at large scales like a country or geographic region.

The probability of deforestation at any given point depends on the probability of deforestation in its neighboring cells. Nearest-neighbor methods look for training samples near a given point to indicate the number of adjacent entities. In short, it is observing patterns in the data and using them to predict new observations.

The statistics behind many of the methods used in this example, such as Markov Chain Monte Carlo (MCMC) methods, involve advanced concepts beyond the focus of this book, but if you're interested in learning more, see the Appendix.

For those who know some statistics, briefly, the Intrinsic Conditional Auto-Regressive (iCAR) model allows estimations of varying probability of deforestation regardless of the nature of immeasurability. For example, it is impossible to measure each individual tree and its impact on the tree cover. We use a logistic regression model of a binary outcome: 1 if a forest pixel is deforested (fewer pixels) and 0 if it is not.

Run the code in the notebook if you are interested in the model preparation or variable selection models. The deep statistical tangent is out of the scope of this book, but understanding the process is quite interesting.[4]

Let's look at the model summary:

```
Binomial logistic regression with iCAR process
  Model: I(1 - fcc23) + trial ~ 1 + scale(altitude) + scale(slope)
  + scale(dist_defor) + scale(dist_edge) + scale(dist_road) + scale(dist_town)
  + scale(dist_river) + cell
  Posteriors:
                      Mean       Std    CI_low    CI_high
       Intercept     -3.84     0.224     -4.23      -3.27
  scale(altitude)     -0.5     0.105    -0.679     -0.293
     scale(slope)   -0.0159    0.0545    -0.117     0.0906
```

---

4 For a deeper dive into the statistical methods at work here, please see: Vieilledent, Ghislain, Vancutsem, Christelle, Bourgoin, Clément, Ploton, Pierre, Verley, Philippe, and Achard, Frédéric. 2022. "Spatial Scenario of Tropical Deforestation and Carbon Emissions for the 21st Century." BioRxiv preprint. *https://www.biorxiv.org/content/biorxiv/early/2022/07/23/2022.03.22.485306.full.pdf*

scale(dist_defor)	-2.06	0.274	-2.51	-1.51
scale(dist_edge)	-6.89	0.44	-7.78	-6.2
scale(dist_road)	-0.0408	0.0573	-0.159	0.0702
scale(dist_town)	-0.0916	0.0444	-0.175	0.0032
scale(dist_river)	-0.0122	0.0347	-0.0838	0.0607
Vrho	3.12	0.852	1.83	5.07
Deviance	1.52e+04	48	1.52e+04	1.54e+04

What you are observing is that when the coefficient is negative, the variable has a negative effect on what you are measuring. For example, the likelihood of deforestation is higher when the distance from the road is smaller.

The MCMC Distance Matrix

First, we sample the data, then we prepare, run, and test the model. The model is summarized in Figure 10-20. The statistics behind many of the methods are advanced concepts beyond the focus of this book but are highlighted in the article "Spatial Scenario of Tropical Deforestation and Carbon Emissions for the 21st Century" by Ghislain Vieilledent et al. Briefly, MCMC methods are used in Bayesian inference and are based on techniques used to generate random sampling sequences to approximate a probability distribution.

MCMC tools, in brief, verify if an algorithm-generated sample is sufficient to serve as an approximation of the at-large distribution. Is your sample size large enough? To simplify a discussion on traces and posteriors, MCMC draws samples from posterior distributions, summaries of uncertainty that you can update based on new information. In a nutshell, once you see the data, you are likely to update your knowledge.

The main purpose is to make sure that the effective size of your sample is not too small and that the samples are similar to the target population. Compare the trace to the shape of the sample histograms to the normal distributions in Figure 10-20. Do you notice any potential skew?

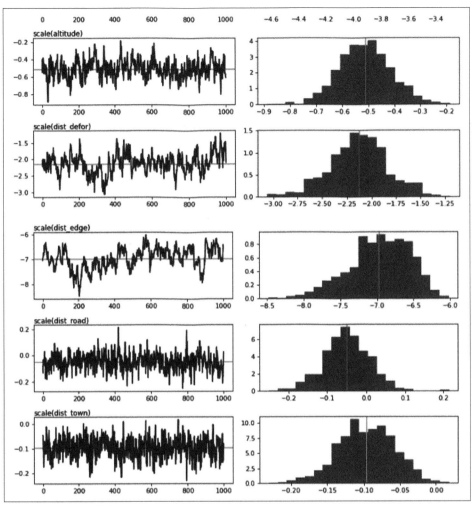

Figure 10-20. MCMC traces

Modeling Deforestation Probability with predict_raster_binomial_iCAR

Using `predict_raster_binomial_iCAR`, you can predict the spatial probability of deforestation from the `model_binomial_iCAR` model by computing block by block over large geographical areas. Selecting just 10 rows should help limit memory drain.

Run the code in the notebooks, and let's look at the generated graphics here.

The output shown in Figure 10-21 shows deforested areas as white and forested areas as black.

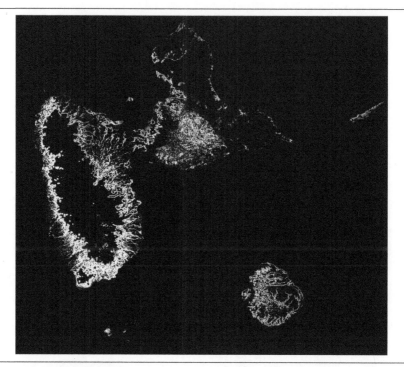

Figure 10-21. Probability TIFF

This model predicts that annual deforestation between 2010 and 2020 was 498.375 hectares per year. For comparison, that is the equivalent of more than 11,000 NBA basketball courts or 930 football fields (including the endzones). The notebook includes code to analyze predictions for future forest-cover change. Figure 10-22 shows historical forest-cover change for the period between 2000 and 2020.

Modeling data and generating probabilities and projections are powerful tools for open source communities tasked with monitoring the threat to our rain forests and the deleterious downstream impact on our ecosystems and climate. Run the code to try modeling the data forward to 2050 and 2100:

```
# Projected forest cover change (2020-2050)
fcc_2050 = far.plot.fcc("output/fcc_2050.tif",
                        maxpixels=1e8,
                        borders="data/ctry_PROJ.shp",
                        linewidth=0.2,
                        output_file="output/fcc_2050.png",
                        figsize=(5, 4), dpi=800)
```

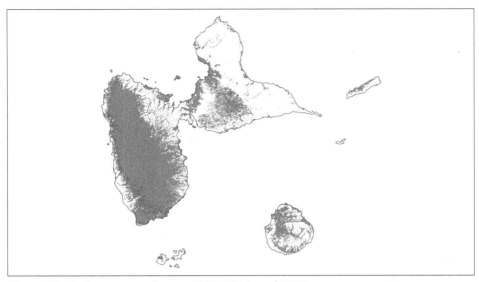

Figure 10-22. Forest-cover change, 2000, 2010, and 2020

The output is shown in Figure 10-23:

```
// Palette for fcc_2050 and fcc_2100
var pal_fcc_proj = [
rgb(227, 26, 28), // 0. Deforestation, red
rgb(34, 139, 34), // 1. Remaining forest, green
];
```

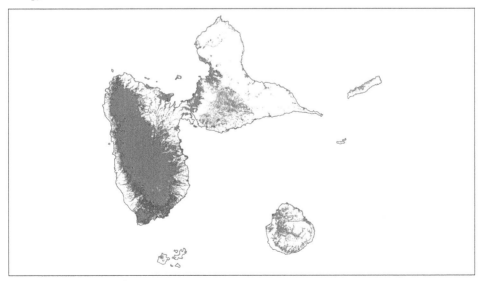

Figure 10-23. Future forest cover, 2050

The persistent impact of ongoing deforestation can be monitored by tracking carbon emissions as well.

Carbon Emissions

The carbon emissions of current and future deforestation can be calculated from the aboveground biomass (*AGB.tif*) file and the function `.emissions()`:

```
# Create dataframe
dpast = ["2020"]
dpast.extend(dates_fut)
C_df = pd.DataFrame({"date": dpast, "C": np.repeat(-99, ndates_fut + 1)},
                    columns=["date","C"])
# Loop on date
for i in range(ndates_fut):
    carbon = far.emissions(input_stocks="data/emissions/AGB.tif",
                           input_forest="output/fcc_" + dates_fut[i] + ".tif")
    C_df.loc[C_df["date"]==dates_fut[i], ["C"]] = carbon
# Past emissions
carbon = far.emissions(input_stocks="data/emissions/AGB.tif",
                       input_forest="data/fcc23.tif")
C_df.loc[C_df["date"]==dpast[0], ["C"]] = carbon
# Save dataframe
C_df.to_csv("output/C_emissions.csv", header=True, index=False)
```

To view the dataframe of deforestation rate estimates:

```
print(C_df)
```

Estimating the persistent rate of change in deforestation rates, annual carbon emissions will continue to increase, as depicted in the output of the code cell:

```
      date    C
0     2020    85954
1     2030    99463
2     2035    152047
3     2040    202883
4     2050    290920
5     2055    335097
6     2060    382887
7     2070    481913
8     2080    587963
9     2085    645071
10    2090    705914
11    2100    844066
```

The model predicts that, in the absence of robust preservation or conservation in the area, carbon emissions will continue to increase rapidly.

Analysis

This section showed you an approach for using your Python and geospatial analytics skills to model and forecast tropical deforestation at the country level. This approach allows us to estimate the individual effect of each of several different environmental factors on the probability of deforestation.

The findings are easy to interpret: deforestation becomes more likely as forested land becomes more accessible. Deforestation is less likely inside protected areas. We also showed that there is a great deal of spatial variability in the deforestation process, and we need to account for that to be able to forecast deforestation in a realistic way at the national scale.

Summary

This chapter provided a glimpse into the available packages you can use to explore and analyze climate change and its potential risks using location intelligence and geospatial tools.

In this book, you have been introduced to geospatial analysis and important Python libraries, packages, and tools. In the beginning, introductory packages like pandas and GeoPandas introduced you to working with dataframes and GeoDataFrames. Tools like NumPy, matplotlib, and Plotly began to appear as important dependencies for more complex packages.

I hope the long journey has been worth your time, and that you'll continue to explore these functions, tools, packages, and more. Any work in data science is iterative. You gain confidence from practice—and, dare I say, from navigating the hiccups along the way.

Kudos on your journey.

Additional Resources

In addition to the GitHub repo for this book, here are some libraries, course, websites, and organizations you may also find helpful.

Python Libraries for Geospatial Analysis

- ArcGIS API for Python (*https://oreil.ly/9YVfx*)
- Census (*https://oreil.ly/VCYPH*)
- Geopandas (*https://geopandas.org*)
- GDAL (*https://oreil.ly/8p4A0*)
- geemap (*https://geemap.org*)
- Forest at Risk (*https://oreil.ly/jYvg9*)
- Missingno (*https://oreil.ly/4yYOg*)
- Xarray (*https://oreil.ly/kb2NC*)
- Osmnx (*https://oreil.ly/tKpqS*)
- Matplotlib (*https://matplotlib.org*) (built on NumPy (*https://numpy.org*))

Resources for Further Exploration

- National Geospatial Program (*https://oreil.ly/cLNOo*)
- Geographic Resources Analysis Support System (GRASS) (*https://grass.osgeo.org*)
- OpenGeoportal (*http://opengeoportal.io*)
- ServirGlobal (*https://oreil.ly/SyE9k*)
- DIVA-GIS (*https://oreil.ly/oTYcL*)

- Natural Earth (*https://www.naturalearthdata.com*)
- OpenStreetMap (*https://www.openstreetmap.org*)
- National Geospatial Digital Archive (*http://www.ngda.org/home.html*)
- National Historical Geographic Information System (NHGIS) (*https://www.nhgis.org*)
- Geography and American Community Survey (US Census) (*https://oreil.ly/mW0Z9*)
- Center of Excellence for Geospatial Information Science (CEGIS) (*https://oreil.ly/EusJr*)
- Python Foundation for Spatial Analysis (Full Course Material) (*https://oreil.ly/XHNfY*)
- Become an Expert, Master Data Science with Python (*https://www.dunderdata.com*)
- PyQGIS 101 (*https://oreil.ly/mz4TS*)
- DataSF (*https://datasf.org/opendata*)
- Data.ny.gov (*https://www.ny.gov/programs/open-ny*)
- GeoPandas 101 (*https://oreil.ly/ZxsPV*)
- NetCDF (*https://oreil.ly/mZWww*)
- Brazil Data Cube (*https://oreil.ly/Gt7cN*)

Bibliography

Bilogur, Aleksey. "Missingno: A Missing Data Visualization Suite." *Journal of Open Source Software* 3, no. 22 (February 2018): 547. *https://doi.org/10.21105/joss.00547.*

Boeing, Geoff. "OSMnx: New Methods for Acquiring, Constructing, Analyzing, and Visualizing Complex Street Networks." *Computers, Environment and Urban Systems* 65 (September 2017): 126–139. *https://doi.org/10.1016/j.compenvurbsys.2017.05.004.*

Chakraborty, T., & Lee, X. "A Simplified Urban-Extent Algorithm to Characterize Surface Urban Heat Islands on a Global Scale and Examine Vegetation Control on Their Spatiotemporal Variability." *International Journal of Applied Earth Observation and Geoinformation* 74 (February 2019): 269–280. *https://doi.org/10.1016/j.jag.2018.09.015.*

Goodman, Cooper, Nathanael Rosenheim, Wayne Day, Donghwan Gu, and Jayasaree Korukonda. *Population Distribution Workflow Using Census API in Jupyter Notebook: Dynamic Map of Census Tracts in Boone County, KY, 2000.* July 2020. Inter-university Consortium for Political and Social Research. *https://doi.org/10.3886/E120382V1.*

Gott, J. Richard III, David M. Goldberg, and Robert J. Vanderbei. "Flat Maps that Improve on the Winkel Tripel." Preprint, submitted February 15, 2021. *https://doi.org/10.48550/arXiv.2102.08176.*

Graser, Anita. "PyQGIS 101: Chaining Processing Tools." *Free and Open Source GIS Ramblings* (blog). *https://anitagraser.com/pyqgis-101-introduction-to-qgis-python-programming-for-non-programmers/pyqgis101-chaining-processing-tools.*

Gray, Jim. "eScience: A Transformed Scientific Method." Presented to the National Resource Council Computer Science Telecommunications Board, Mountain View, CA, January 11, 2007. *http://itre.cis.upenn.edu/myl/JimGrayOnE-Science.pdf.*

Hagberg, Aric A., Daniel A. Schult, and Pieter J. Swart. "Exploring Network Structure, Dynamics, and Function Using NetworkX." Presented at the 7th Annual Python in Science Conference, Pasadena, CA, August 2008. *https://conference.scipy.org/proceedings/SciPy2008/paper_2.*

Mitchell, Bruce, and Juan Franco. *HOLC "Redlining" Maps: The Persistent Structure of Segregation and Economic Inequality*. National Community Reinvestment Coalition. March 20, 2018. *https://ncrc.org/holc*.

National Research Council. *Learning to Think Spatially*(Washington, D.C.: The National Academies Press, 2006). https://doi.org/10.17226/11019.

Open Source Geospatial Foundation. *GDAL/OGR Geospatial Data Abstraction Software Library*. *https://doi.org/10.5281/zenodo.5884351*.

Tobler, W. R. "A Computer Movie Simulating Urban Growth in the Detroit Region." *Economic Geography* 46 (June 1970): 234–240. *https://doi.org/10.2307/143141*.

Vancutsem, C., F. Achard, J.-F. Pekel, G. Vieilledent, S. Carboni, D. Simonetti, J. Gallego, L.E.O.C. Aragão, and R. Nasi. "Long-Term (1990–2019) Monitoring of Forest Cover Changes in the Humid Tropics." *Science Advances* 7, no. 10 (March 2021): 1–22. *https://doi.org/10.1126/sciadv.abe1603*.

Vieilledent, Ghislain. "forestatrisk: a Python Package for Modelling and Forecasting Deforestation in the Tropics." *Journal of Open Source Software* 6, no. 59 (March 2021): 2975. *https://doi.org/10.21105/joss.02975*.

Wu, Qiusheng. "geemap: A Python Package for Interactive Mapping with Google Earth Engine." *Journal of Open Source Software* 5, no. 51 (July 2020): 2305. *https://doi.org/10.21105/joss.02305*.

Wu, Qiusheng, Charles R. Lane, Xuecao Li, Kaiguang Zhao, Yuyu Zhou, Nicolas Clinton, Ben DeVries, Heather E. Golden, and Megan W. Lang. "Integrating LiDAR Data and Multi-Temporal Aerial Imagery to Map Wetland Inundation Dynamics Using Google Earth Engine." *Remote Sensing of Environment* 228 (2019): 1-13. *https://doi.org/10.1016/j.rse.2019.04.015*.

Index

C

Copernicus Open Access Hub (European Space Agency), 67, 204
correlation plots, 237-237
CRS (coordinate reference system), 109
    Cartesian, 181
    checking for geography file and census data, 161
    geometries transforming to, 111
    setting for QGIS project, 29
    specifying for targeting with gdal_warp function, 190
cylindrical projections, 11

D

data cleaning, 165-184
    checking for missing data, 165-175
        data types, 169
        examining summary statistics, 171-175
        metadata, 169
        nulls and non-nulls, 169
        replacing missing values, 175
        uploading to Colab, 166-168
    mapping patterns, 178-183
        shapefiles, 180-183
        visualizing NYPD filtered crime data on a map, 178
    visualizing data with Missingno, 175-178
data dictionaries, 16
    for NYPD Complaint dataset, 167
data models (spatial), 15
data profiles, 157, 158
data questions, asking, 5-7
Data Source Manager (QGIS)
    creating WFS connection, 52
    uploading files with, 25
data types, 171
    examining for each column in your dataset, 169
dataframes, 138
    converting spaces to underscores in, 167
    df.describe function, 171
    df.head function, 139, 179
    df.loc function, 182
    df.shape function, 139
    df.tail function, 139
    GeoDataFrame, 143
    getting acquisition date, 139
    NYPD Complaint dataset, 167
    NYPD.isna function, 174

summary of NYPD.info function, 170
using to create interactive display in deforestation study, 225
datasets
    available from GeoPandas, 146
    census data, 152
    evaluating and inspecting, 16
    exploring data resources, 24
    NYPD_Complaint_Data_Historic, 165
    Xarray, 213
        properties of, 213
datetime object (Python), 133, 169
    transforming timeline data stores as list of dates to, 227
deforestation and carbon emissions in Amazon rain forest using WTSS series (example), 220-229
    analysis, 224
    creating your map, 223-224
    refinements, 225-229
        making your map interactive, 225
        reducing cloud coverage with masking, 226
    setup, 221
        creating your environment, 223
        obtaining the data, 221
deforestation in Guadeloupe, modeling and forecasting with Forest at Risk, 229-245
    analysis, 245
    carbon emissions, 244
    correlation plots, 237-237
    MCMC Distance Matrix, 240
    modeling deforestation probability with predict_raster_binomial_iCAR, 241
    modeling probability of deforestation with iCAR model, 239
    plotting the data, 233
    sampling the data, 235-237
    setup, 230-233
        creating your environment, 231
        downloading and importing packages, 232
        downloading and importing the data, 232
degradation, 220
democratizing data, 2-5
dendrograms, 177
digital elevation SRTM data in EarthExplorer, 201

integration into spatial analysis, 1
iterators, working with, 56-59
magic functions, 105
specifying version for geospatial environment, 72
writing code in QGIS Python Console, 33
Python Console in QGIS, 21, 32-34, 44
discovering attributes in layers, 54
installing, 23
population data visualization, 32
running Python code to add data to the canvas, 58
using processing algorithms in, 59-65
Python plug-in for QGIS, 44-45

Q

QGIS, 20-24
adding basemaps to, 21-24
exploring data resources, 24
exploring the workspace, tree cover and inequality in San Francisco, 42-51
accessing the data, 46-48
addressing the research question, 50
Python plug-in, 44-45
sample data visible in QGIS map canvas, 43
working with Layer panels, 48
gdalinfo displayed in, 191
GDL command line scripts in, 185
installing, 20-21
PyQGIS Python API, 41
using processing algorithms in Python Console, 59-65
visualizing environmental complaints in New York City, 24-30
uploading data to QGIS, 25-29
using Query Editor to filter data, 29-30
visualizing population data, 31-39
loading raster layer, 34
redlining, mapping inequalities, 36-39
working with QuickOSM, 117
QGIS GUI
algorithm input and output parameters in GUI output, 64
algorithms in, 64
QgisInterface class, 42, 44
QgsProcessingRegistry, 61
QuickOSM, working with in QGIS, 117

R

raster data, 15-16
capturing input raster bands using command line and GDAL, 191
changing raster projection with gdal_warp function, 190
exploring in Spyder IDE, 194
exploring open source raster files, 200-206
Copernicus Open Access Hub, 204
Google Earth Engine, 204-206
USGS EarthExplorer, 200
FAR data as GeoTiff raster file, 234
information available about in GDAL, 189
loading raster layer in QGIS, 34
raster functions, 127-131
best for viewing vegetation, NDVI function, 131
generating list for map using Landsat multispectral data, 130
stretch, 135
redlining, or mapping inequalities, 36-39
relations (OSM), 94
return statements, 80

S

scaling factors, 78
seasonal grouping of precipitation data, 219
Sentinel 2 data, 190
shape of data, 169
shapefiles (.shp), 13
for census tracts, 154
mapping data patterns from, 180-183
styling by HOLC grade, 37
uploading to EarthExplorer, 200
Shapely library (Python), 180
shortest distance between points, 101
shutil package, 232
Shuttle Radar Topography Mission (SRTM), 202
sinusoidal projection, 10
skewed distributions, 172
socioeconomics of neighborhoods, 3
spaces, converting to underscores (_), 167
spatial algorithms, 68
spatial analysis, 7, 67
essential facilities, 19-39
exploring spatial data in QGIS, 20-30
visualizing population data, 31-39
spatial attributes, 68

versions, displaying for census packages, 153

W

walkability, 107
warp function (GDAL), 190
ways (OSM), 94
Web Feature Service (WFS), 52
    identifying environmental threats in Massa-
        chusetts, 51-59
        accessing the data, 52-53
        discovering attributes, 54
        styling layers, 59
        working with iterators, 56-59
Web Mercator coordinate system, 10
web scenes, 125
Web Time Series Service (WTSS), 209, 220
    installing and setting up, 223
well-known ID (WKID), 138
well-known text (WKT) markup language, 145
WFS (see Web Feature Service)
WFS/OGC API, 52
!wget call, 154

WGS84 projection, 31
Winkel-Tripel projection, 10
WTSS (see Web Time Series Service)

X

Xarray package (Python), 209
    combining precipitation datasets for 2015
        and 2021, 213-217
    working in, 211-213
    yearly precipitation in continental US by
        month, 218

Z

zero correlation, 177
zero-based indexing, Python data access based
    on, 133
ZipFile module, 232
zoom levels, 136
    in geemap, 75
    preset in Leafmap, 88

About the Author

Dr. Bonny P McClain is a geospatial analyst and self described human geographer / social anthropologist. Dr. McClain applies advanced data analytics—including data engineering and geo-enrichment—to poverty, race, and gender discussions. Her research targets judgments about structural determinants, racial equity, and elements of intersectionality to illuminate the confluence of metrics contributing to poverty. Her work shows that moving beyond ZIP codes to explore apportioned socioeconomic data based on underlying population data leads to discovering novel variables based on location to build more context to complex data questions.

Bonny is a member of the National Press Club, 500 Women Scientists, and The Urban and Regional Information Systems Association (URISA), as well as a former member of the Tableau Speaker Bureau and Investigative Reporters and Editors (IRE), all of which have allowed access to a wide variety of health policy and health economic discussions.

Colophon

The animal on the cover of *Python for Geospatial Data Analysis* is a diamond python (*Morelia spilota spilota*). This snake is a subspecies of carpet python found in Southeastern Australia. They live near the coast of Victoria and New South Wales in Australia, as well as bushland, forests, heaths, and urban and suburban areas. Their climbing and camouflage skills also make it easy for these to adapt to living in the attics, wall spaces, rafters, and roofs of people's homes.

Diamond pythons are medium to large snakes that can grow nearly 10 feet long and live up to 20 years. The species is named for the diamond-shaped rosette pattern of smooth, shiny scales along its back. Diamond pythons are mostly black or dark olive green with a few light spots, or bright yellow with a few dark edges. They are not venomous but constrict their prey, including mice, rats, possums, birds, lizards, and frogs, and swallow it whole. Diamond pythons breed during September and October. Females can lay 54 eggs at a time and guard them over the 55-day incubation period.

While adult diamond pythons have few predators, their young can fall victim to birds of prey, cats, dogs, other snakes, and lizards. These snakes are also threatened by habitat loss. The IUCN lists carpet pythons as a species of "Least Concern"; they do not have a separate status for the diamond python subspecies. Many of the animals on O'Reilly covers are endangered; all of them are important to the world.

The cover illustration is by Jose Marzan, based on an antique line engraving from *Brehms Thierleben*. The cover fonts are Gilroy Semibold and Guardian Sans. The text font is Adobe Minion Pro; the heading font is Adobe Myriad Condensed; and the code font is Dalton Maag's Ubuntu Mono.